SIMPLICIUS

On Aristotle Physics 8.1-5

Translated by István Bodnár, Michael Chase and Michael Share

BLOOMSBURY
LONDON • NEW DELHI • NEW YORK • SYDNEY

Bloomsbury Academic
An imprint of Bloomsbury Publishing Plc

50 Bedford Square
London
WC1B 3DP
UK

1385 Broadway
New York
NY 10018
USA

www.bloomsbury.com

Bloomsbury is a registered trade mark of Bloomsbury Publishing Plc

First published in 2012 by Bristol Classical Press
Paperback edition first published 2014

British Library Cataloguing-in-Publication Data
A catalogue record for this book is available from the British Library.

ISBN HB: 978-1-7809-3210-1
PB: 978-1-4725-3917-5
ePDF: 978-1-4725-0179-0

Library of Congress Cataloging-in-Publication Data
A catalog record for this book is available from the Library of Congress.

The present translations have been made possible by generous and imaginative funding from the following sources: the National Endowment for the Humanities, Division of Research Programs, an independent federal agency of the USA; the Leverhulme Trust; the British Academy; the Jowett Copyright Trustees; the Royal Society (UK); Centro Internazionale A. Beltrame di Storia dello Spazio e del Tempo (Padua); Mario Mignucci; Liverpool University; the Leventis Foundation; the Arts and Humanities Research Council; Gresham College; the Esmée Fairbairn Charitable Trust; the Henry Brown Trust; Mr and Mrs N. Egon; the Netherlands Organisation for Scientific Research (NOW/GW); the Hungarian National Fund for Scientific Research (OTKA); the Ashdown Trust; Dr Victoria Solomonides, the Cultural Attaché of the Greek Embassy in London. The editor wishes to thank Peter Lautner, Barrie Fleet, James Wilberding, Pantelis Golitsis, Michael Griffin, Donald Russell, Michael Share, Richard McKirahan, and Katharine O'Reilly for their comments, Annie Hewitt, David Robertson, Riin Sirkel and Ian Crystal for indexes. He is also grateful to Ian Crystal for preparing the volume for press and to Deborah Blake, the publisher responsible for every volume since the first.

Typeset by Ray Davies

Contents

Preface

Richard Sorabji

In these chapters of Book 8 of the *Physics*, Aristotle (384-322 BC) paves the way for introducing, at the end of Book 8, the Prime Mover of the heavens, which along with most Greeks Aristotle took to rotate around the earth. In Chapter 1, he urges against Democritus that an explanation of motion is needed, even if, as he argues in Chapter 2, motion is eternal, despite (Chapter 3) some non-motion. In Chapter 4,[1] he argues that a body's movement must be caused *by* something distinct from itself, contrary to Plato's view that soul is *self*-moving. Taking the particular case of animals, which are popularly thought of as self-moving, Aristotle says that it is really the soul that moves an animal's body (8.4, 254b27-33, developed in 8.6, 258a1-b9), along with stimuli from the environment (*to periekhon*, 8.2, 253a7-20; 8.6, 259b1-20). But what about the natural motion of air and fire upwards and of earth or water downwards? In *Physics* 2.1, he had distinguished natural motion from the motion of artefacts, by saying that natural objects have an *internal* source of movement, their nature, whereas artefacts have to be propelled from *outside*. But in *Physics* 8.4, a body's inner nature seems insufficiently distinct from it to satisfy the requirement that what is moved should be moved *by* something other than itself. So he introduces a point not made explicit before, that the inner nature of earth, air, fire and water is a source not of causing motion, but of undergoing it (*paskhein*, 255b30-1) at the hands of some *further* agent. To serve as this further agent, he introduces two low-grade causes which do not maintain continuing contact with what they move. To take steam rising from a kettle as an example of rising air, the low-grade agent could be either the person who boils the water in the kettle, or the person who releases the obstacle to its rising by taking the lid off the kettle (256a1-2). The obstacle remover, however, is an agent only accidentally, like the wall off which a ball ricochets. This second agent does not interest Alexander, and Philoponus was later to rule out the obstacle-remover altogether (*in Phys.* 195,24-32). The inner nature still answers the question why (*dia ti*) light and heavy natural objects move to their natural places. But the further agents answer the question *by* what the movement is initiated. If in all motion, even that of animals and of natural rise and fall, a body is moved by something distinct from itself, then the way has been paved for inferring by induction that the heavens too must be moved by *something* distinct from themselves. The induction is not required to show that that something acts in the same way as the celestial mover.

In 8.5, Aristotle argues that, on pain of an infinite regress, we must avoid saying that everything moved is moved by something that is moved in its turn. But he still regards Plato's idea of self-motion as not yet ruled out, and offers a more general argument that in so-called self-movers, we have to find a part that causes and a part that undergoes motion. Hence the prime mover that stops the regress needs to be *unmoved*. Rather unexpectedly, he urges that the soul itself is unmoved, a view that he will defend in his later *On the Soul* 2.5, by saying that when the capacities that in his view constitute the soul are actualised, such actualisation is not to be thought of as an ordinary case of being moved or affected. The reference to soul as unmoved may seem risky, because he does not want the conclusion that the heavenly bodies are moved simply by their own souls as unmoved movers. But eventually in 8.10, without mentioning the issue, Aristotle will argue that the unmoved mover required cannot be a capacity housed in or possessed by the finite body of his universe (as the souls of the heavenly bodies would need to be), because a finite body could not accommodate the infinite power needed for the eternal rotation of the heavens.

The commentary on Aristotle's *Physics* 8.1-5 written by the Neoplatonist Simplicius some time after AD 529 is particularly interesting for two reasons. First, it contains a substantial excursus attacking Philoponus, and this is the subject of Michael Chase's Introduction to Simplicius' response to Philoponus' attack on Aristotle *Physics* 8.1. Simplicius' excursus includes precious fragments of Book 6 of Philoponus' *Against Aristotle*. These fragments, numbered 121-33, have already been included in fourteen pages of Christian Wildberg's 1987 translation of Philoponus' attack on Aristotle in the series Ancient Commentators on Aristotle, and have been further discussed by him elsewhere.[2] Philoponus' attack includes the famous arguments that, on principles accepted ever since Aristotle, the universe cannot have finished going right through a more than finite number of days, much less be going to exceed a more than finite number. These arguments by Philoponus have drawn attention repeatedly and were highlighted again in 1983.[3] The present translation makes more accessible Simplicius' side of the case. This was examined in the first of several writings by Philippe Hoffman, in a collection of 1987.[4] The collection's second edition of 2010 provides a survey and bibliography of writing on Philoponus since 1987, including writing on this subject,[5] and Michael Chase below takes the discussion of Simplicius further.

Secondly, Simplicius reports and quotes extensively the lost commentary of Alexander, holder of the Aristotelian chair in Athens at or soon after AD 200. This is an ideal time to consider Simplicius' use of Alexander,[6] because Marwan Rashed has at this moment published and discussed his newly discovered fragments of Alexander's lost commentary on Books 4 to 8, more on Book 8 than on any other book.[7] In an earlier article, he found Simplicius ascribing to Alexander the opposite of his real view by leaving out of his quotation six words that survive in the newly recovered text.[8] But the overall picture that emerges from comparing the new fragments as a whole is that Simplicius tends to report or quote Alexander accurately and to respect his judgement, even though on certain types of issue he goes on to disagree. The subjects on which Simplicius disagrees suggest to Rashed Simplicius' purposes in writing a commentary.[9] Simplicius saw some Presocratic philosophers, along with Plato, Aristotle and the *Chaldaean Oracles* as constituting a coherent and eternal philosophy, in which Aristotle legitimately discussed the perceptible world, but Plato above all and others emphasised its dependence on a higher intelligible world. Simplicius disagrees with Alexander when the latter interprets Aristotle in a way that rules out the higher causes and that turns merely perspectival differences among the authorities into doctrinal differences. Alexander, he thinks, is of much value because of his mastery of Aristotle's text, but because of these other tendencies, his is the last interpretation of Aristotle outside pagan Neoplatonism that Simplicius can tolerate. Christianity, on this view, is the foe, represented at its worst in Simplicius' Christian contemporary Philoponus. The latter has no understanding of perspectival differences, but seeks to show dissension and error in the eternal philosophy.

If this is right, it is no accident that the two most distinctive aspects of the present text are its attack on Philoponus, and its intensive use, with caveats, of Alexander. For these will have been the two issues which most exercised Simplicius.

Notes

1. This account of Chapters 4 and 5 is chiefly based on that in my *Matter, Space and Motion*, London 1988, Chicago 2006, ch. 13.

2. Christian Wildberg, 'Prolegomena to the study of Philoponus' *contra Aristotelem*' in Richard Sorabji (ed.), *Philoponus and the Rejection of Aristotelian Science*, London 1983, 2nd edn, Supplement 103 to the Bulletin of the Institute of Classical Studies, London 2010, and in *John Philoponus' Criticism of Aristotle's Theory of Aether*, Berlin 1968.

3. Richard Sorabji, *Time, Creation and the Continuum*, London 1983, Chicago 2006, ch. 14, 'Infinity arguments in favour of a beginning'.

4. Philippe Hoffmann, 'Simplicius' polemics', in Richard Sorabji (ed.), *Philoponus and the Rejection of Aristotelian Science*, London 1983, 2nd edn, Supplement 103 to the Bulletin of the Institute of Classical Studies, London 2010.

5. The second edition edition adds, besides the survey and bibliography of the literature on Philoponus since 1987, an interpretation of the excavated

Alexandrian lecture rooms where Philoponus taught. I regret that I omitted reference to Clemens Scholten's *Antike Naturphilosophie und Christlicher Kosmologie in der Schrift 'De Opificio Mundi' des Johannes Philoponos*, Berlin 1996-7, and a series of books and papers by Edward Watts. This was corrected, and Watts' work discussed, in 'Waiting for Philoponus', an introduction to the translation of Aeneas of Gaza and Zacharias of Mytilene, in the series Ancient Commentators on Aristotle, 2012.

6. I have not seen Stephen Menn's paper in preparation on this subject.

7. Alexandre d'Aphrodise, *Commentaire perdu à la* Physique *d'Aristote (Livres IV-VIII)*. Les scholies byzantines, Berlin/New York 2011.

8. Marwan Rashed, 'A "new" text of Alexander on the soul's motion', in Richard Sorabji (ed.), *Aristotle and After*, Supplement 68 to the Bulletin of the Institute of Classical Studies, London 1997, 181-95.

9. Ibid., 23-9.

Conventions

[…] Square brackets enclose words or phrases that have been added to the translation or the lemmata for purposes of clarity, as well as those portions of the lemmata which are not quoted by Simplicius.

<…> Angle brackets enclose conjectures relating to the Greek text, i.e. additions to the transmitted text deriving from parallel sources and editorial conjecture, and transposition of words or phrases. Accompanying notes provide further details.

(…) Round brackets, besides being used for ordinary parentheses, contain transliterated Greek words and Bekker page references to the Aristotelian text.

Abbreviations

Bernabé = A. Bernabé, *Poetarum epicorum Graecorum testimonia et fragmenta*, vol. 1, Leipzig 1987.

Graham = D.W. Graham, *Aristotle,* Physics, *Book VIII,* translated with a commentary, Oxford 1999.

FdV = Hermann Diels and Walther Kranz (eds), *Die Fragmente der Vorsokratiker. Griechisch und Deutsch,* 6. Auflage, 3 Bände, Berlin 1951 (1st edn 1903).

Hardie and Gaye = R.P. Hardie, and R.K. Gaye (trs), 'Physics', in J. Barnes (ed.), *The Complete Works of Aristotle. The Revised Oxford Translation,* vol. 1, Princeton 1984.

Hope = R. Hope (tr.), *Aristotle's* Physics, Lincoln 1961.

Lampe = G.W.H. Lampe (ed.), *A Patristic Greek Lexicon,* Oxford 1961.

LSJ = H.G. Liddell and R. Scott, *A Greek-English Lexicon,* rev. H. Jones; with a new supplement, Oxford 1996.

Ross = W.D. Ross, *Aristotle's* Physics, a revised text with introduction and commentary, Oxford 1936.

SVF = Hans Friedrich August von Arnim, *Stoicorum Veterum Fragmenta,* 4 vols, Leipzig 1903-24.

TLG = *Thesaurus Linguae Graecae* [CD-ROM].

Wehrli = F. Wehrli, *Die Schule des Aristoteles,* vol. 8, *Eudemus von Rhodos,* 2nd edn, Basel; Stuttgart 1969.

Wicksteed and Cornford = P.H. Wicksteed and F.M. Cornford, *Aristotle,* The Physics, with an English translation, vol. 2, London 1968.

Departures from Diels' Text

1121,21: Rejecting Diels' insertion of *aei*.
1122,14: Read *hupetithonto* instead of *hupethento* (Diels, in app. crit.).
1125,14: Read *apodeixis* with MS F, instead of Diels' *apostasis*.
1133,21: Read <*ouk*> *epharmosei* (Wildberg).
1138,2: Replace Diels' comma by a full stop.
1138,20: Read *energeiai* (with iota subscript) instead of Diels' *energeia*.
1139,5: Read *teleiotera tôn allôn* instead of Diels' *teleiotata tôn allôn*.
1143,29: Modify Diels' punctuation, restoring the missing quotation marks.
1148,12: Supply <*epi*> *tôn pros ti* at line 12 (anonymous reader).
1150,33: Read *phthorai* (with iota subscript) instead of Diels' *phthora* (anonymous reader).
1151,7: Read *monou*, with F, instead of Diels' *ekeinou*.
1151,24-5: Read *anankê prohuparkhein kai ekeino* (MS F) instead of Diels' *anankê prohuparkhein, kai ekeinou*.
1153, 8: Read *ho de khronos ê̱* (*hê* Diels) *kinêsis estin ê arithmos kinêseôs*.
1152,12: Delete Diels' comma after *kinêsis*.
1157,13: Diels' *peri tou nun* at 1157, 13 is a mistake for *peri tou nou* (Wildberg).
1157,13: Delete *ei*.
1159,13-14: Diels' *ho ho khronos* is, of course, a typographical error.
1159,29: Read *pantos* with MS F, instead of Diels' *pantôs*.
1168,27: Read *eikhe* instead of *hexei* (Diels, in app. crit.).
1170,4: Punctuate Diels' text as follows : *kai ei men phtharta, kan phthareiê, pasa phthora dia kinêseôs* <sc. *estin*> *ktl*.
1170,23: Read *alla* instead of Diels' *allo*.
1170,36: Read *hôste kai* for Diels' *hôste tai*.
1171,11: Read *marainomenês* for Diels' *manainomenês*.
1174,22-5: Alter Diels' punctuation, removing his question mark at line 22 and inserting one at line 25.
1174,35: Read *ton kosmon* with MS F, instead of Diels' *einai*.
1174,37: Read *hoti* instead of *to* (Diels, in app. crit.)
1178,32: Read *hoti *hou* pro tês geneseôs* etc. instead of Diels' *ou* (Wildberg).
1179,11: omit *mê*.
1180,5: Read *prosêgage* instead of Diels' *proêgage*.

1184,32: Read *ê* instead of Diels' *kai*.
1193,15: Changing both instances of *kineisthai* to *kineitai*.
1194,12-13: Deleting *tôi sôritêi*.
1195,10: Adding *ou* before *tinos*.
1196,4: Punctuating with a comma rather than a full stop after *akinêton*.
1198,12: Punctuating with a comma rather than a full stop after *akousai*.
1198,13 : Punctuating with a comma rather than a semicolon after *esti*.
1198,23: Reading *eite* rather than *ei te*.
1199,3-4: Changing *genomenon* to *genomenê*.
1200,37: Punctuating with a comma rather than a full stop after *kinêsin*.
1202,1: Punctuating with a comma rather than a semi-colon after *phthora*.
1202,19: Adding *hoion* before *leukos*.
1205,28: Changing the second instance of *mê on* to *kenon*.
1205,33: Changing *pepisteumenon* to *pepistômenon*.
1211,27: Read *kinêtikon* instead of the *kinein* of the manuscripts.
1217,23: Read *enekhthênai* instead of the typo *anekhthênai*.
1217,32: Read *metabalontôn* instead of the *metaballontôn* of the manuscripts.
1218, 9: Read *metaballonta* instead of the *metaballein* of the manuscripts.
1224,9: Changing *periousan* to *periiousan*.
1225,4: Punctuating with a comma rather than a full stop after *prôtou*.
1225,24: Reading *kinei to* and *katho* rather than *kinoito* and *kath' hauto*.
1225,28: Changing *tôi* to *mê*.
1225,29: Reading *to te* rather than *tote*.
1226,1: Punctuating with a full stop after *kinêsis*.
1227,14-15: Changing *kinoun ti* to *kinounti*.
1230,10: Punctuating with a full stop rather than a comma after *dedeiktai*.
1230,13: Punctuating with a full stop rather than a comma after *dunamei*.
1230,27: Changing *kinoumena* to *kinounta*.
1230,31: Omitting *ê* before *heauto*.
1231,28-9: Removing the brackets around *adunaton … sundramein* and punctuating with a colon after *eipôn* and a comma rather than a full stop after *sundramein*.
1232,37: Supply *mallon* after *autokinêton* (Diels, in app. crit.).
1236,36: Read <*to kath'*> *hauto* instead of *auto* at line 36 (R. Sorabji).
1237,34: Omit *prôtên*.
1246,18: I conjecture *autos enistamenos tou apo tou merous* <*tou autokinountos to autokinêton kineisthai, kai to holon ktl*>.
1247,46: Delete *akinêton* after *to kinoun*.

Simplicius' response to Philoponus' attacks on Aristotle's *Physics* 8.1

Michael Chase

The section devoted to *Physics* 8.1 is one of the most extensive and interesting in Simplicius' commentary on *Physics* 8. On the one hand, it contains Simplicius' usual meticulous comments on the text of Aristotle, who here begins his demonstration of the eternity of motion. As is his wont, the Stagirite starts out with a critical survey of the views of his predecessors, which gives Simplicius the opportunity to quote and explain a number of important fragments of Presocratic philosophers (Anaximander, Anaximenes, Heraclitus, the Atomists, Diogenes of Apollonia, and especially Empedocles). But the bulk of Simplicius' commentary on *Physics* 8.1 consists of one of his famous digressions, in which he quotes and attempts to refute several fragments from Book 6 of *Against Aristotle on the Eternity of the World*, written by his Christian rival John Philoponus some time in the 530s. Many of the arguments of both Philoponus and Simplicius concerning time, eternity, and the nature of the infinite, are of considerable philosophical importance, as a number of recent studies have shown. Quite apart from the intrinsic interest of the various arguments mobilised by both interlocutors, however, Book 8.1 of Simplicius' *in Physica*, together with his *Commentary on the de Caelo*, provide us with vitally important documents concerning the conflict between pagans and Christians in the second quarter of the sixth century AD.

The mid-sixth century was an interesting period in the history of philosophy. By this time, the triumph of Christianity was pretty well complete in the Roman empire, where it had been the official religion, if not since the time of Constantine I, then certainly since 380 under Theodosius I. In 529, the emperor Justinian sealed the fate of pagan philosophical education by ordering the closure of the Platonic Academy at Athens, forbidding pagans to teach anywhere within the empire.[1]

Simplicius and Philoponus

Although they seem never to have met, Simplicius and John Philoponus both began their philosophical studies at Alexandria under Ammonius, who taught there between 475 and 526 AD.[2] But while Simplicius soon left for Athens, Philoponus remained at Alexandria, first writing fairly standard commentaries on Aristotle, based on the notes he took at Ammonius' classes.[3] It was precisely in 529, however, the year of Justinian's edict, that Philoponus[4] began to publish treatises in which he defended a Christian viewpoint, criticising the doctrines of pagan philosophers.[5] He began with a work entitled *On the Eternity of the World against Proclus*, in which he refuted the arguments in favor of the world's eternity by Proclus, the great Athens-based teacher of Ammonius.[6] Philoponus' treatise entitled *Against Aristotle on the Eternity of the World*, which Simplicius sets out to refute in his commentaries on Aristotle's *de Caelo* and *Physics*, is somewhat later, and was probably written in the 530s.[7] It seems likely that Philoponus' choice was not unconnected with what was happening at Athens: perhaps, as some Arabic sources state, Philoponus felt the need to distance his Neoplatonism from pagan philosophy, and point out that its doctrines could, after all, be reconciled with Christianity. The Islamic philosopher al-Fârâbî (*c.* 870-950), who wrote at least one refutation of Philoponus' arguments,[8] remarked in this context:

> One may suspect that his intention from what he does in refuting Aristotle is either to defend the opinions laid down in his own religion about the world, or to remove from himself the suspicion that he disagrees with the position held by the people of his religion and approved by their rulers, so as not to suffer the same fate as Socrates.

Al-Fârâbî thus has two explanations, complementary rather than alternative, for Philoponus' decision to turn against Aristotle. Both could be characterised as socio-ideological. Philoponus felt pressure to conform to Christian beliefs,[9] and so he set out to refute Aristotle's pagan world-view, either because he sincerely believed his own Christian views were correct and Aristotle was wrong, or because he was afraid for his own safety unless he was perceived to support the Christian rather than the pagan view.[10] The first view is more likely, given that we now know that Philoponus was indeed a convinced Monophysite Christian, spending the last part of his life composing Christian theological treatises, some of which, ironically enough, served only to get him condemned for the heresy of tritheism on 3 January 568.[11]

Philoponus, *Contra Aristotelem*

In the *Contra Aristotelem*, Philoponus set about refuting Aristotle's views on the eternity or perpetuity of the world. As a Christian, Philoponus felt obliged to defend the Biblical account, according to which God created the world from nothing in six days, some six thousand years previously. Philoponus' treatise is lost, but the fragments that remain, preserved mainly by Simplicius,[12] show that it consisted of eight books. In the first five, Philoponus attacks Aristotle's views on the nature and existence of a fifth element, the so-called ether, eternally moving in a circle, as set forth in his *de Caelo*, Book 1, 2-4, with a digression on *Meteorology* 1.3. In the sixth book, in which we are interested here, Philoponus attacked Aristotle's arguments in *Physics* 8.1 in favour of the eternity, or rather the perpetuity, of motion, time, and therefore the world. According to Philoponus, the world as a whole was created at a specific moment in time and will also be destroyed at a subsequent moment, to be replaced by a superior world. Such doctrines are anathema to Simplicius, as we shall see.

Simplicius on Philoponus

When we start to read that part of Simplicius' commentary on *Physics* 8 in which he reports Philoponus' objections against Aristotle,[13] it is immediately clear that Simplicius does not like Philoponus very much. He never refers to him by name, but usually as *houtos* (this guy), or as the Grammarian. He also calls him a Telchine, one of the mythological blacksmiths and magicians of Rhodes who, by Late Antiquity, had become synonymous with backbiters or slanderers. Philoponus is a jaybird, or a barking dog, and his arguments are 'heaps of garbage', or filth: Simplicius calls upon Heracles to divert the river Alpheus to clean out the excrement that his arguments have caused to accumulate in the minds of his readers. By constantly emphasising that Philoponus is a grammarian (Greek *grammatikos*), Simplicius is able to emphasise that his opponent is not even a professional philosopher, but a mere teacher of literature, a greenhorn who has a superficial acquaintance with some notions of philosophy. For Simplicius, Philoponus is an *opsimathês*,[14] someone who comes to learning late in life, which implies that he was probably somewhat younger than Simplicius.[15] The vast length of his writings, claims Simplicius, is intended to dazzle the layman, even though much of his material is plagiarised from Alexander of Aphrodisias and Themistius.[16] His intended audience is, moreover, made up of dim-witted students and morons in general.[17] In short, according to Simplicius, Philoponus is uneducated, superficial, thick-witted, and he writes like someone who is insane, drunk, and maniacal.

For Simplicius, these upstart Christians, the worst of whom is the sophist Philoponus, blaspheme against the heavens, eliminating the difference in substance between the celestial and sublunar worlds.[18] In so doing, they ignore the passages in their own holy Scriptures, which teach that 'The heavens proclaim the glory of God, and the firmament announces his handiwork' (Psalm 18). Incomprehensibly, they consider filthy, corruptible matter, such as that of corpses (i.e. the relics of martyrs), to be more worthy of honour than the heavens. As for Philoponus, he dares to proclaim that the light emanating from the heavens is no different from the light emitted by glow-worms and fish-scales. For Simplicius, in contrast, to denigrate the heavens is to blaspheme against the Demiurge, who brought them into being, he whom the late Neoplatonists identified with Zeus or the Intellect.[19]

Above all, Simplicius despises Philoponus and his co-religionists because of their anthropomorphic conception of God. Since Philoponus believes God is like a human being, it is only natural that he thinks God's production, the heavens and the world as a whole, will perish as the works of human beings do. But as Philippe Hoffmann emphasises,[20] taking God to be an individual is:

> a radical inversion of the philosophical attitude, which consists in rising *above* individual humanity.

Hoffmann goes on to quote H.-D. Saffrey,[21] who writes that in Neoplatonism:

> … man is nothing; particular, individual man is nothing but the degradation of Man with a capital M. (…) Man's misfortune is to be an individual, and the entire effort of philosophy is directed to raising oneself back up to the universal and the All.

By anthropomorphising their God, moreover, the Christians are guilty of making him arbitrary and capricious. When Philoponus (*Against Aristotle*, fr. 120 Wildberg) suggests that God may have created the elements in the beginning, then handed over their subsequent administration to Nature (rather like the Newtonian concept of a God who winds up the celestial clockwork, and then leaves it to run on its own), Simplicius is, as usual, scandalised:[22]

> Who in his right mind could conceive of such a change in God, such that not having created earlier, in the briefest moment of time he should become the creator of the elements alone, and then cease from creating once again, handing over to Nature the generation of the elements out of one another, and of the other things from the elements?

What shocks Simplicius here is the arbitrariness attributed to God. He is said to create the world: fine, says Simplicius, although it would require a long argument to agree on the sense of 'create' that is appropriate here. But why should he have decided to create at one moment rather than another?[23] And why should he then stop creating, like some factory worker clocking in and out of the plant? Like Leibniz some 1200 years later, Simplicius cannot tolerate the idea that God's behaviour might be arbitrary or capricious, that is, that he might act without having a *sufficient reason* for acting in the time, place, and way he did. Simplicius' own Neoplatonic doctrine of emanation escapes this particular problem (although it is less successful in avoiding others):[24] emanation, he argues, can be considered as a continuous creation,[25] one that has no beginning or end, so that there is no room for asking: why did God create six thousand years ago, rather than seven thousand?

As Hoffmann has shown, Simplicius' attitude toward Philoponus and his co-religionists is symptomatic of the general attitude of educated pagans at the end of Antiquity toward Christians. The Christians are an impious group of atheists and revolutionaries, whose only redeeming virtue is that they will not be around for long: their doctrines will soon wither away, like the gardens of Adonis. In their desire for glory, they are like Herostratus of Ephesus, who burned down the temple of Artemis in 356 BC, just because he wanted to be famous. Motivated by the search for glory rather than the pursuit of truth, they have failed to purify their rational soul, with the result that they allow themselves to be motivated by their passions and imagination rather than reason.

Aristotle, *Physics* 8.1

As is well known, Aristotle maintained that time and motion were not created but eternal, as was the world: no matter what moment in time, or what motion in physical space you choose, there will always have been a moment or motion before it, and there will always be one after it. In this sense, because there is no first or last moment of the world's existence, the world is eternal. Plato's position was harder to pin down. In his most famous and influential dialogue, the *Timaeus*, he talks as though a creative divinity, which he calls the Demiurge or craftsman, created the world, time, and the human soul, fashioning them out of a chaotic hodgepodge of wildly moving elements, or rather proto-elements.[26] Yet Plato had presented this account in the form of a myth, and there was considerable debate in Antiquity over whether it should be understood literally, or merely in some kind of a symbolic or allegorical way.[27]

The eighth book of Aristotle's *Physics*, which some interpreters, such as al-Fârâbî, considered the culmination of the entire book,[28] sets out to prove the existence of an unmoved Prime Mover, responsible for all the motion in the universe. To accomplish this, Aristotle starts out in *Physics* 8.1 by trying to prove that motion is eternal, time is eternal, and therefore the entire world as a whole is eternal.

To prove that motion is eternal, Aristotle starts out from the definition of motion he had already given in *Physics* 3.1, 210a10ff.: Motion is the actuality (*energeia* or *entelekheia*) of what is movable insofar as it is movable. This, Aristotle claims, implies that before motion can take place, the things that are capable of motion must already exist. But these things are either generated, or eternal. If generated, their existence must be preceded by the motion or change that generated them; if they are eternal, but were not *always* in motion, then they must have begun to move at a specific point in time, prior to which they were at rest. But if this is the case, since rest is the privation in motion, then while they were at rest there must have been some cause that kept them at rest. Before these things begin their motion, therefore, there must have been another change or motion that overcame the cause that was maintaining them at rest. Aristotle's conclusion is that no matter whether the things capable of motion are generated or eternal, there is always a change or motion previous to any change or motion one chooses to consider. In this sense, then, motion is eternal. There is no such thing as a first motion.

Aristotle's second argument is based on his definition of time as the number of motion according to the before and after (*Physics* 4.10-12). Since time is the number of motion, if there is always time, there is always motion as well. Aristotle therefore (*Physics* 8.1, 251b10ff.) goes on to give a series of arguments for the eternity of time.[29]

Aristotle's first argument for the eternity of time is from authority: all natural philosophers except Plato, he says, have agreed that time is eternal. The key point here is that Aristotle takes the account of creation in Plato's *Timaeus* quite literally.

Aristotle's second argument for the eternity of time is based on the nature of the present instant or the now (Greek *to nun*). By Aristotle's definition, the now is the end of one period (viz., the past), and the beginning of another one (the future). Since every now thus implies time before and after it, it follows that there can be no first or last now, and hence that time is eternal. Finally, Aristotle goes on to show that these arguments prove that time, and therefore motion, not only had no beginning but will also have no end, for whichever instant or *nun* you consider, there will always be one after it. Time is thus beginningless and endless, infinite *a parte ante* and *a parte post*, as the Latins would say, and as the Arabs would say, both *azalî* and *abadî*.[30]

Simplicius vs. Philoponus: the gloves come off

All kinds interesting issues are raised in the debate between Simplicius and Philoponus over the interpretation of *Physics* 8.1. Among the most interesting of these, from a purely philosophical viewpoint, is Philoponus' attempt to refute Aristotle by arguments based on the nature of infinity; these arguments are the subject of section 9 (a), pp. 175-80 of the second volume of Richard Sorabji's *Philosophy of the Commenta-*

tors. In order to rule out the possibility of beginningless time, Philoponus adduces the fact that, according to Aristotle, there can be no actual infinite; that no infinite series can be traversed or increased; that no one infinite series can be larger than another; and that no infinite quantity can be a multiple of another infinite quantity. These arguments have been amply discussed by Sorabji and others, so I shall merely point here to some less-studied aspects of the debate between Philoponus and Simplicius.

Simplicius on the created nature of Christ

The first example occurs when Simplicius is answering Philoponus' attempt to overturn what he calls the 'famous axiom of the philosophers', to the effect that nothing can be generated (Greek verb *genesthai*, adjective *genêton*) out of nothing, an axiom Philoponus considers essential for the pagan proof that motion is eternal (Simpl. *in Phys.* 1143,20ff.). Philoponus contends that contrary to what Aristotle says, what is generated can indeed come into being out of nothing, or more precisely out of what does not exist in any way (*ek tou medamêi mêdamôs ontos*). For Philoponus, God creates matter, from which it follows that, contrary to what the Pagans claim, not everything that comes into being originates out of what exists (*to on*). Not only matter, moreover, but all forms within matter, and, in short, everything except the First is created, according to Philoponus, with only the First being ungenerated and uncaused.

Simplicius takes advantage of this opportunity to question Philoponus' Christian orthodoxy. He first cites Aristotle at *Physics* 1.8, 191a24ff., who argues that nothing can be generated out of nothing, but that whatever comes into being must do so out of its own privation. This allows Simplicius to make fun of Philoponus for not understanding what the philosophers mean by 'generation' (Greek *genesis*): it is not, as the Grammarian thinks, what depends on just any kind of cause, but 'what has been assigned its passage to being within a part of time'. But now Simplicius (*in Phys.*, p. 1144, 28-32 Diels) administers the *coup de grâce*:

> And since [Philoponus] says that only the First is ungenerated and without a cause, joining [the epithet] 'without a cause' to [the epithet] 'ungenerated', he also says, *not even showing respect for those who share his views*, that what comes after the First *is also generated and is created*. For he too says that what is generated is created (…)

If Philoponus affirms that everything after the First – that is, presumably, God the Father – is created, then that includes Christ the Son. But to say that Christ is created (Greek *poiêthen*) goes against the Nicene Creed, which held that Christ was 'begotten, not made':[31] it is, in fact nothing other than Arianism.[32] I find it quite surprising, – and I'm not aware that it has been noticed before – that the resolutely pagan Simplicius should be apparently so up to date on the niceties of theoretical Christology.

Simplicius and Philoponus on perpetuity (*aidiotês*)

Another example of Simplicus' attacks on Philoponus' Christian faith comes in the context of Aristotle's 'proof' that time is everlasting, based on the fact that all his predecessors, except for Plato, said it is:

> But so far as time is concerned we see that all with one exception are in agreement in saying that it is uncreated (…) Plato alone asserts the creation of time, saying that it is simultaneous with the world, and that the world came into being (Aristotle, *Physics* 8.1, 251b14-19).

Philoponus provides three counter-arguments.

First, just because five or ten men say time was generated, this is no reason to prefer their testimony to that of Plato. We cannot judge the validity of opinions on the basis of how many people support them; if we did, Aristotle, who was the only one to introduce a fifth element (ether) alongside the four traditional ones of earth, air, fire, and water, would be out of luck. Secondly, Aristotle's claim that we should follow the majority is hard to square with his statement in the *de Caelo* (1.10, 279b12ff.) that even though all the other natural philosophers say the world is generated, he shows it to be ungenerated. Thirdly, since Plato said in the *Timaeus* that 'time came into being with the heavens', he is more consistent than the others, who claim that the world is generated but that time is ungenerated, although neither can exist without the other. Philoponus goes on to argue that we should not accept the testimony of the natural philosophers that time is ungenerated, since Aristotle says they were wrong in every other respect. Besides, Philoponus says, he could point out many illustrious ancient philosophers who claim that time is generated.

Simplicius begins his refutation of Philoponus' arguments by claiming that Aristotle does not affirm that the testimony of other philosophers is demonstrative proof, but he only quotes them to back up his own demonstrations; such testimony helps to persuade beginners.

Secondly, Simplicius claims that when Plato and Aristotle call the world and time 'generated' (Greek *genêton*), they do not mean the same thing. When Aristotle *seems* to argue against Plato, he is in fact '… arguing not against Plato, but against those who understand the term "generated" according to its surface meaning' (1165,4-5). Simplicius has already explained, at 1154,4ff., that by 'generated' Aristotle means what exists subsequently after having been non-existent,[33] and what exists in a part of time. Plato, in contrast, means by it all that is not true, viz. intelligible, being, and not simultaneously entire: 'generated', according to Plato, means whatever has an external cause of its being.[34] Simplicius continues by claiming that Philoponus is showing his ignorance when he attributes to Plato his own understanding of the term 'generated', viz. that it refers to what comes into existence after having previously been non-existent. Philoponus was, as usual, too dumb to understand Plato when, in the *Timaeus*, he says that the Demiurge wished to make the world as similar as possible to its intelligible model. The model (Greek *paradeigma*), according to Plato, was characterised by eternal everlastingness (*tên aiônion aidiotêta*), and so the Demiurge provided the world with temporal everlastingness (*tên khronikên aidiotêta*) by bestowing upon it time, as an image of eternity.[35]

Simplicius on the Egyptian origins of Genesis

Continuing his refutation of Philoponus, Simplicius denies that Aristotle differs from Plato when, in the *de Caelo*, he introduces a fifth element as characteristic of the heavens. This is a good example of the pagan Neoplatonist concern, more or less universal since the time of Porphyry, if not already of Antiochus of Ascalon, to reconcile Plato and Aristotle.[36] The two great founders of philosophy cannot be allowed to contradict one another. If they sometimes appear to do so – as even the Neoplatonists were obliged to concede – then the reason is, as Simplicius states of the apparent contradiction between Plato's and Aristotle's use of the term 'generated' (*genêton*),

> ... it was the ancient usage to argue against the surface meaning out of consideration for more superficial understandings. Since, then, 'generated' was said of things that having previously not existed, later existed, therefore, arguing against this meaning of the term, Aristotle seems to censure Plato for having said 'generated', but in fact he is censuring not Plato, but those who have attached 'generated' in this sense to time and to the world.

Whereas Aristotle *appears* to say, *expressis verbis*, that Plato was the only one to say that time is generated, and that he was *wrong* to do so, in fact, on the Neoplatonist explanation that Simplicius adopts, Aristotle was criticising not Plato, but those who understood only the superficial or apparent meaning of 'generated', viz. that something begins to exist after having been non-existent. Plato's 'real' meaning, which professors like Simplicius explained to their students, is that to say that a thing is 'generated' actually means that it depends on an external cause for its existence, is not intelligible, and is not a simultaneous whole but has its being in becoming.

Finally (p. 1166,20ff.), Simplicius concludes his refutation of Philoponus on this point with a final argument. I'm not aware of any modern scholarship dealing with this passage from Simplicius, but one notable scholar who did call attention to this passage was Ralph Cudworth, in his *True Intellectual System of the Universe* (I quote from his translation in ch. 4, p. 313 of the 1678 edition):

> (...) Simplicius a zealous Contender for the Worlds Eternity, affirms the Mosaick History of its Creation by God, to have been nothing else but *muthoi Aiguptioi*, Egyptian Fables.
>
> The Place is so considerable, that I shall here set it down in the Authors own Language,
>
> If Grammaticus here means the Lawgiver of the Jews, writing thus, [In the beginning God made Heaven and Earth, and the Earth was invisible and unadorned, and Darkness was upon the Deep, and the Spirit of God moved upon the Water:] and then afterward when he had made Light, and separated the Light from the Darkness, adding [And God called the Light Day, and the Darkness Night, and the Evening and the Morning were the First Day] I say, if Grammaticus thinks this to have been the First Generation and Beginning of Time, I would have him to know that all this is but a Fabulous Tradition, and wholly drawn from Egyptian Fables.

When copying this passage in the manuscript that came to be known as Marcianus Graecus 227, the thirteenth-century scribe Georgios could not restrain his indignation, writing in the margin: 'Behold this dog Simplicius, saying that the words of Moses are myths!'.[37]

Unfortunately, Simplicius does not tell us where he got his information. Some modern scholarship has taken up the hypothesis that Egyptian influence can be discerned in the opening chapters of Genesis.[38] This is particularly the case with the so-called Cosmogony of Hermopolis. It may be, however, that Simplicius is referring not to the ancient religion of the Egyptians, but to the doctrine of such Egyptianising astrologers as Nechepso and Petosiris, who, by casting the 'horoscope of the world', presupposed that they could determine the precise moment of its coming into being.[39]

The posterity of the debate between Simplicius and Philoponus in Arabo-Islamic philosophy

The debate between Simplicius and Philoponus on the question of the eternity of the world had a considerable impact on Medieval Arabic-language philosophy. Several of Philoponus' works were widely known in Arabic,[40] often in several versions, and although it has been an axiom in Islamic studies since the study by H. Gätje[41] that Simplicius' *Com-*

mentary on the Physics was unknown to the Arabs, this view may have to be revised as scholarship continues to progress.

Al-Kindî (*c.* 801-873), often considered the first Islamic philosopher, adopted many of Philoponus' arguments in favour of creation *ex nihilo*[42] and the impossibility of an eternal world, and his views were influential on the *Plotiniana Arabica* and the *Liber de Causis* (probably written in the second half of the ninth century),[43] al-Ghazâlî (*c.* 1058-1111)[44] and the entire Kalâm. Probably in reaction against Kindî, al-Fârâbî (c. 872-951) wrote refutations of Philoponus' *Against Aristotle*:[45] in particular, his lost work *On Changing Beings* seems to have been devoted to refuting Philoponus' arguments and arguing in favour of the world's eternity.[46] Al-Fârâbî's views were transmitted with more or less complete approval by the Andalusian philosopher Ibn Bâjja (Avempace, *c.* 1095-1198), adhered to more ambivalently by Ibn Sînâ (Avicenna, *c.* 980-1037),[47] while the situation in Ibn Rushd[48] (Averroes, *c.* 1126-1198) is even more complex. Initially favourable to al-Fârâbî's views,[49] Ibn Rushd later repudiated most of them after lengthy study and meditation, a process we are now in a position to trace thanks to meticulous research on the various phases of composition of his commentaries on the *Physics*.[50]

This research increasingly shows a number of close parallels between the commentaries on the *Physics* of Simplicius and of Ibn Rushd.[51] Philoponus had sought to deny Aristotle's view that motion requires the previous existence of objects capable of that motion, by adducing the case of the elements, which move toward their natural places as soon as they come into existence. If it should be argued against Philoponus that wood pre-exists fire's upward motion, he will argue, on the one hand, that such a change is generation rather than motion; on the other, he will deny that fire can be the entelechy of wood. If it were, then wood would have two contrary entelechies (downward motion *qua* wood, upward motion *qua* fire), and two contraries (fire and wood) would have the same entelechy. Simplicius responds by invoking the notion of the twofold nature of potentiality: fire's upward motion is a case of the kind of power, potentiality, or capacity (Greek *dunamis*) that is perfect and constantly projects actuality (*energeia*), whereas wood's *dunamis* for upward motion is a case of what tends toward but does not yet possess such perfect *dunamis*. What Philoponus fails to understand, claims Simplicius, is that fire possesses upward motion in the sense of perfect *dunamis*, while wood possesses it in the sense of imperfect *dunamis*.

Averroes so closely reproduces the arguments for both sides of this debate that R. Glasner, for instance, who holds to the orthodox view that Ibn Rushd knew Philoponus' work but not that of Simplicius, is obliged to assume[52] that the Cordoban arrived independently at solutions that closely parallel those of Simplicius. It seems much more economical and plausible, however, to assume that Ibn Rushd and the rest of the *falâsifa* did have access to at least part of Simplicius' commentary on the *Physics*, although the precise mechanism of this transmission will have to await further study.

Notes

1. R. Sorabji, 'Infinity and the creation', in R. Sorabji (ed.), *Philoponus and the Rejection of Aristotelian Science*, London/Ithaca 1987, 164-78, p. 164 (see now the updated edition published by the Institute of Classical Studies, London 2010). The doubts expressed by Alan Cameron about the extent and efficacy of Justinian's edicts are probably ill-founded, cf. Ph. Hoffmann, 'Simplicius' polemics', in Sorabji (ed.) 1987/2010.

2. I. Hadot, 'La vie et l'œuvre de Simplicius d'après des sources grecques et arabes', in eadem (ed.), *Simplicius: sa vie, son oeuvre, sa survie*, 3-39, p. 7.

3. The first redaction of Philoponus *in Phys.* dates from 5 May 517 (L.S.B. MacCoull, 'A new look at the career of John Philoponus', *Journal of Early Christian Studies* 3.1 (1995), 47-60, p. 49). See now Cl. Scholten, *Johannes Philoponos. De aeternitate mundi / Über die Ewigkeit der Welt*, 2 vols, Turnhout: Brepols 2009 (Fontes Christiani 64/1-2), vol. 1, 48-65.

4. It is true that Philoponus' criticism of pagan philosophers begins as early as his *Commentary on the Physics*; see the translations by K. Algra and P. Huby in the current series.

5. This has been questioned by Lang and Macro (*Proclus, On the Eternity of the World* = (*De Aeternitate Mundi*); Greek text with introduction, translation and commentary by Helen S. Lang and A.D. Macro; argument I, translated from the Arabic by Jon McGinnis, Berkeley 2001), who affirm that the *De Aeternitate Mundi* is a philosophical work bereft of Christian apologetics, and that, in general, 'there is virtually a complete absence of evidence for a Christian commitment in Philoponus' philosophical writings'. This claim seems absurd, and has been persuasively refuted by M. Share (tr.), *Philoponus: Against Proclus On the Eternity of the World 1-5*, London/Ithaca 2004 (Ancient Commentators on Aristotle), pp. 4ff.: cf. Scholten 2009, 70ff.

6. MacCoull ('A new look at the career of John Philoponus', p. 48) refers to this work as containing Philoponus' 'pathbreaking rejection of the steady-state universe in favour of a "Big-Bang Theory" consistent with the Christian doctrine of Creation'.

7. R. Sorabji in C. Wildberg (tr.), *Philoponus: Against Aristotle On the Eternity of the World*, London/Ithaca 1987 (Ancient Commentators on Aristotle), p. 24.

8. Al-Fârâbî, *Against John the Grammarian*, 4, 8, p. 257 Mahdi. For a critical evaluation of this testimony, see U.W. Lang, *John Philoponus and the Controversies over Chalcedon in the Sixth Century: a study and translation of the 'Arbiter'*, Leuven 2001 (Spicilegium sacrum Lovaniense: études et documents; 47), p. 7f. Ironically, Philoponus himself (*Aet. Mundi* 9,4, 331,20-5 Rabe) suggests that when Plato calls the world created by the Demiurge a 'happy god' (*eudaimôn theos*), he was merely yielding to popular superstition, lest he should suffer the same fate as Socrates. Cf. K. Verrycken, 'Philoponus' interpretation of Plato's cosmogony', *DSTradF* 8 (1997), 269-318, p. 278.

9. This is basically the view of K. Verrycken, 'The development of Philoponus' thought and its chronology', in R. Sorabji (ed.), *Aristotle Transformed*, London 1990, 233-74.

10. cf. H. Chadwick, 'Philoponus the Christian theologian', in R. Sorabji (ed.), *Philoponus and the Rejection of Aristotelian Science*, 41-56, p. 42: '...

Philoponus saw the Athens affair as an opportunity and a challenge, whether he wrote in order to attract Justinian's favour by an attack on the principal architect of late Neoplatonic dogmatics or to avert unwelcome attention from the Alexandrian philosophers by demonstrating that not all of them were motivated by a cold hatred of Christianity as Proclus was'. Some ancient sources claim Philoponus wrote his anti-Aristotelian works in order to make money; cf. Verrycken 'The development of Philoponus' thought', pp. 258-63. MacCoull ('A new look ...', p. 52), for her part, explains al-Fârâbî's report in the context of contemporary conflicts between Monophysite and Chalcedonian Christians.

11. cf. Chadwick, 'Philoponus the Christian theologian'. More precisely, Philoponus 'broke away from the miaphysite communion and endured anathema for the sake of his rather abstruse Trinitarian doctrine' (Lang, *John Philoponus*, p. 8). On the theological background of Philoponus' Trinitarian works, see also L.B.S. MacCoull, 'The historical context of John Philoponus' *De Opificio Mundi* in the culture of Byzantine-Coptic Egypt', *Zeitschrift für Antikes Christentum* 9.2 (2005) 397-423, p. 412ff. Philoponus was condemned a second time in 575, this time for his unorthodox views on the nature of the resurrection body, and again at the Third Council of Constantinople in 680-1.

12. The fragments have been translated in this series by Christian Wildberg, *Philoponus: Against Aristotle on the Eternity of the World*, London/Ithaca 1987.

13. Simplicius, *in Phys.* 1130,1ff. Diels.

14. ibid., 1133,10.

15. Philoponus was probably born in Egypt around 490 (MacCoull 'A new look ...', p. 49), and died around 575 (eadem, 'The historical context ...', p. 415).

16. Simplicius, loc. cit., 1130,5.

17. *anoêtôn anthrôpôn* 1130,1. C. Scholten (*Johannes Philoponus, De Opificio Mundi = Über die Erschaffung der Welt*, Freiburg etc. 1997, 3. vols [Fontes Christiani Bd. 23/1-3], vol. 1, p. 14) suggests Philoponus may have written his *De Opificio Mundi c.* 557 in order to prove that Christians were not simpletons who deserved the derision of their pagan colleagues. It has also been suggested (R. Walzer, *Greek into Arabic. Essays on Islamic Philosophy*, Harvard 1962 (Oriental Studies, 1), 195; E. Behler, *Die Ewigkeit der Welt. Problemgeschichtliche Untersuchungen zu den Kontroversen um Weltanfang und Weltunendlichkeit in der arabischen und jüdischen Philosophie des Mittelalters*, München/Paderborn/Wien 1965, p. 132) that this work was a response to critiques from the Christian side, who complained that Philoponus had not made enough use of Scripture in his previous polemical works.

18. This was, of course, one of the aspects of Philoponus' thought that was appealing to, and influential upon, Galileo; cf. M. Rashed, 'The problem of the composition of the heavens (529-1610): a new fragment of Philoponus and its readers', in P. Adamson et al. (eds), *Philosophy, Science and Exegesis in Greek, Latin and Arabic Commentaries*, vol.1 (Bulletin of the Institute of Classical Studies supplement 83.1), London 2004.

19. On the question of the identity and ontological rank of the Demiurge, see R. Sorabji, *The Philosophy of the Commentators, 200-600 AD: 400 years of transition, a sourcebook*, 3 vols, London 2004, vol. II, §8 (e), pp. 170-3, with further references. See also M. Chase, 'What does Porphyry mean by *theôn patêr*?', *Dionysius* 22, Dec. 2004, 77-94, pp. 88ff.

20. 'Simplicius' polemics', cited above, n.1.

21. H.-D. Saffrey, 'Théologie et anthropologie d'après quelques préfaces de Proclus', in *Images of Man in Ancient and Medieval Thought (Studia Gerardo Verbeke ... dicata)*, Louvain 1976, p. 208.

22. Simplicius, loc. cit., 1147,1ff.

23. cf. Sorabji, *Philosophy of the Commentators*, vol. 2, §9(b), pp. 180-1.

24. In particular, proponents of continuous creation are often accused of eliminating God's freedom of will, and making the creative process necessary and automatic. Against this view, see, for instance, K. Kremer, '*Bonum est diffusivum sui*. Ein Beitrag zum Verhältnis von Neuplatonismus und Christentum', *ANRW* II.36.2 (1987), 994-1032.

25. cf. Sorabji, *Philosophy of the Commentators*, vol. 2, §8(f), pp. 173-4.

26. This is how Aristotle interprets the cosmogony of the *Timaeus*, cf. *On the Heavens* 1, 10-12.

27. cf. W. Wieland, 'Die Ewigkeit der Welt (Der Streit zwischen Joannes Philoponus und Simplicius)', in D. Heinrich et al. (eds), *Die Gegenwart der Griechen im neueren Denken. Festschrift für H. G. Gadamer zum 60. Geburtstag*, Tübingen 1960, 219-316, p. 293; R. Sorabji, *Time, Creation and the Continuum: Theories in Antiquity and the Early Middle Ages*, London/Ithaca 1983, pp. 268-72; M. Baltes, *Die Weltentstehung des platonischen Timaios nach den antiken Interpreten, Teil I*, Leiden 1976 (Philosophia antiqua 30); L. Judson, 'God or nature? Philoponus on generability and perishabilty', in R. Sorabji (ed.), *Philoponus and the Rejection of Aristotelian Science*, 179-96, p. 179.

28. A short treatise by al-Fârâbî against Philoponus' arguments is extant (cf. Mahdi, 'Alfarabi against Philoponus', *Journal of Near Eastern Studies* 26.4 (1967), 233-60), but it concerns only Philoponus' arguments against the *de Caelo*. It has been suggested that al-Fârâbî's lost treatise *On Changing Beings* (*Fî al-mawjûdât al-mutaghayyira*) was devoted to Philoponus' arguments against *Physics* 8; cf. H.A. Davidson, 'John Philoponus as a source of medieval Islamic and Jewish proofs of creation', *Journal of the Ancient Oriental Society* 89 (1969), 357-91, p. 360; Rashed, 'Al-Fârâbî's lost treatise ...', passim.

29. As Simplicius explains (1152,24ff.), Aristotle uses the following hypothetical syllogism: if time is everlasting, then motion is everlasting. But the antecedent is true, therefore, so is the consequent.

30. On this terminology, cf. J. Jolivet, *Perspectives arabes et médiévales*, Paris Vrin, 2006 (Études de philosophie médiévale, LXXIX), p. 224ff.

31. 'I believe in one Lord, Jesus Christ, the only Son of God, eternally begotten (*gennêthenta*) of the Father God from God, light from light, true God from true God, begotten, not made (*gennêthenta ou poiêthenta*), of one Being with the Father; through him all things were made'. On the distinction between *poiêma*, which is outside of God's essence and created when God so wills, and *gennêma*, a characteristic feature of God's essence that arises eternally, independently of his will, see M. Wacht, *Aeneas von Gaza als Apologet. Seine Kosmologie im Verhältnis zum Platonismus*, Bonn 1969 (Theophaneia 21), p. 92ff.

32. It was also the doctrine of the Gnostic Ebionites, for that matter: cf. Epiphanius, *Panarion, Anacephalaeosis* II, 30, 1.

33. This is the seventh of the seven meanings of 'generated' (Greek *genêtos*) enumerated by Porphyry, who denies it is applicable to Plato. Cf. Baltes, *Weltentstehung*, p. 137ff.

34. These correspond to meanings 3-4 of *genêtos*, as distinguished by the Middle Platonist Calvisius Taurus; cf. Baltes, *Weltentstehung*, pp. 105-21. Porphyry also accepts the interpretation of *genêtos* as meaning *ap' aitias*, 'from a cause', cf. Baltes, loc. cit., 143.

35. On the distinction between eternal and temporal perpetuity, cf. Proclus, *Elements of Theology* prop. 55; *in Tim.* I, 278,4ff. Diehl.

36. On this theme, see G.E. Karamanolis, *Plato and Aristotle in Agreement? Platonists on Aristotle from Antiochus to Porphyry*, Oxford 2006.

37. *ton kuna Simplikion hôde moi skopei phaskonta muthous tous logous Môuseôs*. Cf. Kalbfleisch's Preface to Simplicius, *in Phys.*, *CAG* IX, p. XIV.

38. cf. J.D. Currid, 'An examination of the Egyptian background of the Genesis cosmology', *Biblische Zeitschrift* 35 (1991), 18-40.

39. On this astrological notion of the *thema* or *genitura mundi*, cf. Macrobius, *Commentary on Scipio's Dream* I, 21; Firmicus Maternus, *Mathesis* III, 1, 1; A. Bouché-Leclercq, *L'astrologie grecque*, Paris 1899, 185-8; 498ff.; F. Boll, *Sphaera*, Leipzig 1903, 234; 373; W. and H. Gundel, 'Planeten', *RE* XXIV.2 (1950), 2149-51; R. Beck, *Beck on Mithraism, Collected Works with New Essays*, Aldershot 2004, pp. 70, 90-2; 162; 178-9; 222-3; 288.

40. Scholten 2009, 199ff.; E. Gannagé, 'Philopon (Jean-). Tradition arabe', in R. Goulet (ed.), *Dictionnaire des Philosophes Antiques. 5, de Paccius à Rutilius Rufus. A, de Paccius à Plotin*, Paris 2011, pp. 503-63. These works include the *Against Proclus*, the *Against Aristotle*, parts at least of the *Commentary on the Physics*, and a monograph, lost in Greek, entitled *Book on the Proof of the Contingency of the World*.

41. H. Gätje, 'Simplikios in der arabischen Überlieferung', *Der Islam*, 59 (1982), p. 6-31. The author is nevertheless obliged to concede (p. 14) that while Averroes never mentions the name of Simplicius, there are numerous parallels in the Cordoban's works with passages from Simplicius' commentaries on the *Physics* and the *De Caelo*.

42. cf. R. Walzer, 'New studies on al-Kindi', *Oriens* 10 (1957) p. 203ff. = *Greek into Arabic. Essays on Islamic Philosophy*, Harvard 1962 (Oriental Studies, 1), 175-205, p. 187ff.; H.A. Davidson, *Proofs for Eternity, Creation and the Existence of God in Medieval Islamic and Jewish Philosophy*, Oxford 1987, p. 106ff.; P. Adamson, *Al-Kindî*, Oxford 2007, p. 74ff.; E. Gannagé, 'Philopon (Jean-). Tradition arabe', pp. 546-8.

43. cf. F. Zimmermann, 'The origins of the so-called *Theology of Aristotle*', in J. Kraye and C.B. Schmitt (eds), *Pseudo-Aristotle in the Middle Ages: The Theology and other Texts,* London 1986 (Warburg Institute Surveys and Texts 11), pp. 110-240; P. Adamson, *The Arabic Plotinus: A Philosophical Study of the Theology of Aristotle*, London 2002; C. D'Ancona, 'La teologia neoplatonica di "Aristotele" e gli inizi della filosofia arabomusulmana', in R. Goulet et al. (eds), *Entre Orient et Occident: la philosophie et la science gréco-romaines dans le monde arabe*, Vandoeuvres-Genève 2010, pp. 1-32 (Entretiens sur l'Antiquité Classique 57).

44. See, e.g., S. Van den Bergh's Introduction to his translation of Averroes' *Tahafut al-Tahafut* (*The Incoherence of the Incoherence*), London 1954, vol. I, p. xviiff.

45. *al-radd 'alâ Yaḥyâ al-naḥwî* (*Against John the Grammarian*); cf. Mahdi, 'Alfarabi against Philoponus'.

46. cf. M. Rashed, 'Al-Fârâbî's lost treatise *On changing beings* and the possibility of a demonstration of the eternity of the world', *Arabic Sciences and Philosophy* 18 (2008), 19-58; 'On the authorship of the treatise *On the harmonization of the opinions of the two sages* attributed to al-Fârâbî', *Arabic Sciences and Philosophy*, 19 (2009), 43-82.

47. cf. J. Janssens, 'Creation and emanation in Ibn Sînâ', *Documenti e Studi sulla Tradizione Filosofica Medievale* 8 (1997), 455-77.

48. cf. R. Glasner, *Averroes' Physics. A Turning Point in Medieval Natural Philosophy*, London 2009.

49. Like Simplicius, Fârâbî assumes that Aristotle's goal in *Phys.* 8.1 is to show that motion is eternal because every motion one may wish to choose is preceded and followed by another. Ibn Rushd begins by accepting this view, but gradually comes to believe that Aristotle is talking not about motion *per se*, but investigating whether the first motion of the first mobile (i.e. the heavens), is caused by the Prime Mover, is eternal or generated in time. Cf. Glasner, *Averroes' Physics*, p. 75ff.

50. In addition to Glasner, see the pioneering work by Josep Puig Montada: 'Averroes and Aquinas on *Physics* VIII 1: a search for the roots of dissent', in *Knowledge and the Sciences in Medieval Philosophy. Proceedings of the Eighth International Congress of Medieval Philosophy (S.I.E.P.M.), Helsinki 24-29 1987*, Helsinki 1990, vol. 2, 307-13; 'Les stades de la philosophie naturelle d'Averroès', *Arabic Sciences and Philosophy*, 7 (1997), 115-37; 'Zur Bewegungsdefinition im VIII. Buch der Physik', in G. Endress and J.A. Aertsen (eds), *Averroes and the Aristotelian Tradition*, Leiden etc. 1999, 145-59; 'Averroes y el problema de la eternidad del movimento', *La ciudad de Dios*, 212 (1999), 231-44.

51. With Simplicius, *in Phys.* 1133,20-1138,11 Diels, for instance, cf. Averroes, *Epit. Phys.* 8, 251b28-252 a6, p. 234ff. Puig; *Long in Phys.* VIII, 1, comm. 4, 340 Iff.; *Long in De Cael.* IV, comm. 24, p. 705, 156-61 Carmody.

52. Glasner, *Averroes' Physics*, p. 91.

SIMPLICIUS
On Aristotle
Physics 8.1-5

Translation

Simplicius the philosopher, commentary on Book Theta of Aristotle's Physics, which is the eighth. 1117,1

Chapter 1

Translated by Michael Chase

The last book of Aristotle's *Physics,* which is the eighth by number of the entire treatise – but the Peripatetics are accustomed to entitle it *Theta*, by alphabetical order – is full of many fine theories. For it puts the best end to the treatise on the physical principles, demonstrating that the first physical movement is everlasting, and clearly proclaiming the everlastingness of the universe by means of its everlasting motion, and, what is greatest, attaching all of physical theory to first philosophy, demonstrating that all physical existence is dependent on the supernatural cause, when [Aristotle] demonstrates that every physical body is in motion, and that all that moves is moved by something. He also reveals that what causes motion in the primary and proper sense of the term [pertains to] the substance that is immobile, always in the same state and identical. Since this is the goal of the book, it would rightly be worthy of all anticipatory zeal.[1] Yet since one of the contemporary backbiters,[2] considering it no great feat to slander men alone, and therefore extending his slander to the very heavens and the entire world, considers their everlastingness problematic, and having partaken, neither successfully nor graciously, of a few 1118,1 false notions taken from philosophy, wrote five lengthy books against what is shown in *On the Heavens* about their perpetuity,[3] whose unsoundness and boldness, directed only to stupid students, I tried to display when explaining the first treatise of *On the Heavens*. He enlisted his sixth book against everlasting motion, which is proved there, and everlasting time, for if they were to remain unrefuted, it was in any case necessary for [the heavens], which move by everlasting motion in everlasting time, to be everlasting as well. It seemed to me appropriate, now that I have come to this book,[4] to display to lovers of learning the counter-arguments that have been raised against it, whatever they may be and from whatever disposition they may have been proposed. Here, too, after the explanation of Aristotle's words, I shall try to put them to the test, by juxtaposing to them the objections against them.

250b11-12 Did motion come into existence at some moment, not existing previously[, and again does it perish in such a way that nothing moves, or did it not come into existence, nor does it perish, but it always existed and always will exist, belonging to beings as immortal and inexhaustible, being like] a kind of life for all things that arise by nature?

That the discourse on motion is necessary for the physicist has already been shown at length. Yet no one would doubt that it pertains to the person discussing motion to inquire whether motion is everlasting, or whether it is such that it came into being at one moment, having previously lacked existence, and again perishes, so that there is no more motion. It is clear, however, that [Aristotle] is not investigating any of the individual sublunary motions (for each of these has a beginning and an end), nor is he here investigating whether motion is everlasting *qua* unified and continuous (he will discover this later), but he is investigating motion in general: whether there was a time when there was not any kind of motion whatsoever, nor was any being in motion, or whether there will be some time when nothing will move; or whether these two alternatives reveal themselves to be impossible, but there always was and always will be motion, so that 'this – that is, motion – is an immortal, inexhaustible property of things that are'.[5] For motion is not one of the things that subsists by itself, but it is one of those that belong to other things, that is, to substances, which are
1119,1 beings in a more proper sense than [accidents], which have their being in [those substances]. This is why he says that motion belongs to beings, not because motion is not-being, but because substances are beings in the proper sense, and it is through them that accidents partake of being. With regard to 'a sort of life as it were to all naturally constituted things',[6] it was said in accordance with Aristotle's exactness, which agrees with Plato, for we say that those things are alive that move from within and by themselves. If, then, all physical things have nature as the principle of motion within them, they too could rightly be said to be alive, in accordance with their inner aptitude for motion. After all, nature is the last of lives;[7] this is why it has its being in a physical body as its substrate, and it is, as it were, more 'alive' that it is a life in the proper sense of the term.[8] Motion in accordance with it is an aptitude for motion, as we shall learn in this book. And [in order to be convinced] that Aristotle uttered this, too, in a manner consonant with Plato, calling internal motion, *qua* physical, 'a sort of life', listen to what is written about internal motion in the tenth [book] of the *Laws*:[9] 'When we see that this motion has shown itself in a thing composed of earth, water or fire, separately or in combination, how should we describe the character resident in such a thing? [Clinias]: Am I right in supposing you ask whether, when the thing moves itself, we speak of it as alive?' But this was said about the life that is self-moved in the proper sense,

viz. psychic life. Aristotle therefore rightly said that physical motion is 'a sort of life', and not simply life, for life in the proper sense has the same ratio to physical motion as the soul has to nature. Alexander,[10] for his part, says that 'a sort of life as it were to all naturally constituted things' was said to indicate that motion is immortal and inexhaustible. For if motion is to physical things as life is to animate things, since it is impossible to be animate without life, so [it is impossible] to be physical without motion. If, then, physical things are everlasting *qua* physical (for they did not begin at any moment, nor will they perish into not-being, since the destruction of physical things is a change into something else that is physical, as when water perishes it becomes air), it is clear that the motion by which they are physical is also everlasting. But perhaps Aristotle did not prove the everlasting nature of motion from the perpetuity of physical things – for so far, physical things have not yet been demonstrated anywhere to be everlasting – but from the very definition of physical motion, as we shall learn, he will prove that physical motion is everlasting, and from this that physical bodies [are everlasting as well]. He said 'to all naturally constituted things' because motion is not like the life of water or air, but as of all physical things 1120,1
qua physical, and *qua* having within them the principle of motion. Therefore, motion taken in the common sense is also everlasting.

250b15-18 For that motion exists, all who have said something about nature [agree, because they create worlds, and because the whole subject of their speculations, for all of them, is generation and perishing, which cannot exist] in the absence of motion.

Since [the statement that] motion is everlasting is significant of what kind of thing it is (*tou hopoion ti*), but the problems both 'what is it?' (*to ti estin*) and 'does it exist at all?' (*to ei estin*) precede 'what kind of thing is it ?',[11] he indicated what motion is in the third book of the treatise, but he does not seem to have shown whether or not it exists. Therefore, proposing here to demonstrate what concerns what a thing is like, he rightly teaches the reason for taking for granted that motion exists. For if all natural philosophers agree with one another that motion exists, although they differ over whether or not it is everlasting, there would be nothing absurd in taking for granted that motion exists. For he proved that neither are those who eliminate motion natural philosophers, nor does argument against them pertain to the natural philosopher, because they destroy the principles of physics. The reason why all natural philosophers discuss motion is 'the construction of the world, and that they all study the question of becoming and perishing'.[12] He says that natural philosophers 'construct the world', not in the sense that they say the world is generated from time, but because they transmit its constitution by means of motion. For in this sense, Democritus constructs worlds, as do Empedocles and Anaxagoras: the former saying that the atomic bodies and the four elements are combined and separated, while Anaxagoras says that the homoiomeries are secreted from the mixture. Yet combination, separation, and secretion are motions. They all speak of generation and destruction, too, which, even if they are not motions, do not take place without motion; for generations and destructions are preceded by local motions, alterations, increases, and diminutions.

> **250b18-23** But whoever says there are infinite worlds[, some of the worlds coming into being, while others perish, state that there is always motion, for their generations and perishings must take place along with motion. Those who say there is one [world] or not always,[13]] make corresponding suppositions with regard to motion as well.

Having provided himself with a justification for assuming that motion
1121,1 exists, viz. the agreement of the natural philosophers on this point, he once again exhibits, from their disagreement over the perpetuity of motion, that the investigation into perpetuity is necessary, saying that some of the natural philosophers assumed that motion is everlasting, some that it is not everlasting, and they adjusted [their doctrines] of motion to their assumptions about the worlds. For some, like the followers of Anaximander,[14] Leucippus, Democritus,[15] and later the followers of Epicurus,[16] assuming the worlds to be infinite in number, also assumed they were generated and destroyed, with some always being generated, and others destroyed, *ad infinitum*. They also said motion is everlasting, for there is no generation or destruction without motion. Of those who say there is only one world, however, those who say it is unengendered in time and imperishable, like Plato and Aristotle, assumed that motion is also everlasting. Those who say there is always a world – not always the same one, but another one always coming into being according at periodic intervals of time – make the one world generated and destructible, as Anaximenes,[17] Heraclitus,[18] Diogenes,[19] and later the Stoics;[20] and it is clear that they hold the same opinion about motion, for whenever there was a world, then motion must necessarily exist. Empedocles, however, if one were to understand him according to the external form of his verse,[21] as saying that sometimes there is the Sphere, sometimes a world, it is clearly because he observed motion in the generation of the world, that he assumed it was always coming into being and perishing. But if he wants both the Sphere and the world to exist always, he considered that there is motion in the world as well.[22] Anaxagoras,[23] Archelaus,[24] and Metrodorus of Chios[25] seem to say that the world had a beginning in time, and they also say that motion had a beginning. For since beings were at rest during the previous time, they say motion came to be within it from the intellect, through which motion the world came into being. They, too,

apparently assume a beginning of the construction of the world, for the sake of pedagogical order. Anaxagoras, for his part, clearly says that intellect brings about the cosmic process of distinction from intelligible unity, in which 'all things were together',[26] as he says. Alexander, however, says that Plato also makes the world subsist from a temporal beginning, except that he said that before the generation of the world, there was also an irregular and disorderly motion among beings. For he says: 'The god took over all that is visible – not at rest, but in irregular and disorderly motion –, and brought it from disorder into order'.[27] But 1122,1
that Plato does not say that the universe came into being from a temporal beginning in the sense that time pre-existed, is clear from the fact that he clearly says that time came into being together with the heavens. Thus, it must not be considered that what is irregular and disorderly pre-existed in time according to him either. Yet in order to show that the generative disorderliness subsists in accordance with the abasement from being and with the order deriving from the Demiurge, which is always setting it in order, he indicated each one separately in his discourse.[28] He did the same thing in the *Statesman*, separating the Demiurge from the world in discourse, and contemplating it as it is carried down 'to the ocean of dissimilarity'.[29] Since this is the division of opinions about the worlds, some, saying they are infinite, in the sense that some are always coming into being while others perish, used to say that motion is everlasting; whereas those who said it is one and always the same, or one but not always the same, made suppositions[30] in a manner appropriate to their doctrine of the world with regard to motion as well. For those who say [the world] is always one and the same also say that motion is everlasting; likewise, those who say that it is one, but different at different times, since the generation of one follows upon the destruction of another, also say that motion is everlasting. For if the world always exists, motion must also always exist. But if, maintaining that [the world] is one, they say that it came into being from some temporal beginning, so that it can truly be said that there was a time where there was no world, and perishing at a certain [moment of] time, it will no longer exist, then motion too will clearly have a beginning and an end. Many, constructing a world, say the one world came into being, wishing to transmit the order of creation in discourse, but we do not know of any natural philosopher who says that the one world comes into being and perishes, so that there is no longer a world.

The letter of the text reading 'those [who say] it is either one or not always'[31] seems to be lacking in conciseness, for 'one' lacks 'and always', while 'not always' is missing 'and one', so that the complete form would be as follows: 'those [who say] that it is either one and always the same, or one, but not always'. This could be divided into two: either in the 1123,1
sense that it is not always the same, but a world always exists, or in the sense that it does not always exist. Yet what was lacking was remedied by this valid induction: whatever they declare about the world, he says,

they make assumptions about motion according to the same account. For if a world exists, coming into being and perishing, then motion must exist, but if it does not exist, nor does it come into being or perish, then there must be no motion.

250b23-251a4 But if it is ever possible that nothing should be in motion[, this must occur in [one of] two ways: either as Anaxagoras says (for he says that after all things were together and at rest for an infinite time, the Intellect impressed motion upon them and separated them), or as Empedocles [says], viz. that [all things] are alternately in motion and at rest: in motion when Love makes the one out of many, or Strife makes many of one, but they are at rest in the intermediary periods. For he says:[32]

Insofar as the one learned to be born out of many,
And again, as the one is divided, the many emerge.
Thus they come into being, and have no abiding life.
But insofar as they never cease their continuous exchange,
so far are they forever immobile in a circle,

For one must suppose that by] 'but insofar as (...) exchange' he means 'from one [state] to another'.[33]

Having said that of natural scientists, some supposed motion to be everlasting, others that it is not everlasting, each group in a manner appropriate to its own opinions about the world, he proposes to eliminate the arguments affirming that it is not everlasting, but that it is such that 'it is possible that at some moment nothing should be in motion'. He sets forth two modes in which it is possible that at some moment nothing moves, and writing the names of their champions Anaxagoras and Empedocles alongside them, it is by eliminating these two modes that he introduces the fact that motion must necessarily be everlasting. There are two such modes, because once motion has appeared, if 'it is possible that at some moment nothing should be in motion', it must be either in the sense that there was no motion previously, as Anaxagoras seemed to say, making clearer the always co-existent subsistence of the second things from the first things by means of the manifestation that seems to be said according to time (which is itself a mode of assumption, even if one later assumes there is no motion. This is why he did not mention it). But Anaxagoras[34] seemed to say that since all things were together and at rest throughout the infinite previous time, the world-making intellect, wishing to separate the forms, which he calls homoiomeries, instilled motion within them. For no corporeal separation could come into being without motion. The second mode[35] is the one according to Empedocles, who produced rest and motion in alternation. For he supposed that both the intelligible and the sensible world consisted of the same four elements, one in the manner of a model, that is, and the other in the manner of an image. The efficient causes were, of the intelligible, Love, which 1124,1 produces the Sphere through unification, which he also calls god (and he sometimes uses the neuter form 'it was a sphere'[36]), and Strife of the sensible world, when it predominates imperfectly, producing this world through separation. Even in this world, one can see both unification and separation, the former in the heavens, which one might rightly call 'sphere' and 'god', and the other in what is beneath the moon, which, since it is primarily what needs to be set in order, and is called 'world'[37] in the more proper sense. Yet both unification and separation can also be observed in the sublunary realm, both of them always [existing], but with a different one at different moments predominating in different parts, or at various times. After all, Empedocles says that here, too, Love and Strife predominate alternately, in the case of men, fish, wild beasts, and birds. He writes as follows:[38]

> Consider the marvellous mass of mortal limbs:
> at one time all coming together into one by Love,
> all the limbs that have been allotted a body,
> at the peak of flourishing life;
> at another time again, dismembered by evil quarrels
> they are scattered, each one alone, along the shore of life.
> So it is too for plants, and fish in their watery palace,
> and the mountain-running beasts, and the wing-sped birds.[39]

Aristotle, however, having cited Empedocles' words, in which he considers the latter to be transmitting [a doctrine concerning] motion and immobility, considers motion according to the generation of the one from the many and of the many from the one, since Empedocles also clearly says:[40]

> Thus they come into being, and have no abiding life.

For it was said earlier that motion accompanies generation, while Empedocles seems to see immobility in the everlasting identity of the mutual change of the one and the many; for such is the meaning of [the verses]:[41]

> But insofar as they never cease their continuous exchange, 1125,1
> so far are they forever immobile in a circle.

Alexander is familiar with this meaning as well, but he says that Aristotle does not take it in this sense, but understands these verses to be about change as well, when [Empedocles] says 'but insofar as they never cease their continuous exchange', that is, they do not cease transforming from these things into those. 'They are also everlasting by this exchange and their mutual change into one another in turn, that is, their change into one another is everlasting. Individually, indeed, none of the things that come into being through Strife and Love from the elements is everlasting, nor is their life abiding, but their change into one another is everlasting. This is what is meant by "are they forever".' Having said this, word for word, Alexander adds: 'then, in addition to this, "immobile in a circle" would be said of these things in ordinary speech', that is, having become immobile in each revolution and each perfection, which he calls 'circle'. But perhaps the demonstration[42] of 'immobile in a circle' is unpersuasive, at the same time as it is unintelligible, for it lacks 'are they forever'. Instead, one should say that Aristotle's understanding is that insofar as they always transform into one another, their transformation and their motion is everlasting, but insofar as they transform from hence to thence and become individual things (*tade*), [becoming] one from many or many from one, for instance, and after the change they sometimes stop with being one, and sometimes with being many, in this sense, being periodically restored, after the change, to the form of the one or of the many, they are immobile throughout that period, until they begin to transform again. Themistius[43] obviously holds to this notion as well, for he says that rest takes place in the times between changes.

> **251a5-8** Yet we must investigate what is the case in these matters[, for it is useful to see the truth not only with a view to speculation about nature, but also with a view to the method of inquiry] concerning the first principle.

He shows how the investigation of everlasting motion is necessary, since it contributes to physical theory (for if physical motion is everlasting, one by continuity and the other by uninterrupted change, it is clear that physical things are everlasting, some by virtue of remaining identical in number, others by the fact that the destruction of one is the coming into being of another), but it also contributes to the ascent toward the first principle, the contemplation of which no longer pertains to the study of physics, but to first philosophy. It also contributes
1126,1 to it in that everlasting motion will be proved to take place by an immobile cause, which is entirely transcendent of physical substance.

> **251a8-16** Let us first begin from what we determined earlier [in the physical books. We say that motion is the actuality of the movable *qua* movable. The things capable of moving must therefore exist in the case of each motion. Indeed, even without the definition of motion, everyone would agree that it is necessary that what is in motion is what is capable of being moved by each

> motion – for instance, what is altered is what is alterable, and what is transported is what is changeable with regard to place – so that there must be something combustible before it is burnt,] and something that can cause burning before it burns something.

Having set out to prove that motion is everlasting by several proofs, he carries out the first from the definition of motion, which he says was defined in the *Physics*, calling 'Physics' the first five books of the *Physics*, which he often calls *On Principles*. For he gave the definition of motion in the third book,[44] from which he now recalls that motion is the actuality of what is movable, *qua* movable, that is, with the object in motion remaining potential as long as it moves. For once it has completely discarded its potentiality and has come to be in act with regard to the [aspect] according to which it was said to be potential, it no longer moves, but comes to rest in actuality. Yet it continues to be moved insofar as it possesses something potential that is in transition to actuality. On the basis of this concept of motion, he proves that motion is everlasting, not only [motion] properly so called, but change, under which generation and destruction are also classified; for this is the definition of such motion. What he proves to be everlasting is not this particular, individual motion, such as this whitening (this is immediately false, for individual motion has a beginning and an end), but motion taken absolutely. Thus, no matter what motion is taken, it is always proved that another motion, whatever it may be, pre-exists, and another one exists after it. It is therefore not possible to take either the initial motion, such that there would be a moment when there was no motion, with nothing moving prior to the first motion, or the last one, so that once again there would be no movement after that one, since nothing is moved. And it is clear that the motion that is everlasting in this sense is neither one, nor continuous, but consecutive. Thus, he assumed from the definition of motion that what is to be moved must previously be movable by nature, and such that it is capable of moving, since all generated actuality proceeds from potentiality, and nothing could come into being that had not been capable of coming into being, nor could anything be moved naturally that did not have the capacity of being moved (for in generated things, the potential is prior in time to what is in actuality, since things that come into being progress from imperfection to perfection). Reminding us in a few words of this, then, from the definition, of the utterance that says 'it is therefore necessary 1127,1
for realities to exist that are capable of moving according to each movement',[45] he did well to observe 'that even without this definition, everyone would agree that it is necessary for that to be moved which is capable of being moved according to each movement'.[46] He exhibits this in the case of every species of motion, saying that 'it is what is capable of alteration that is altered',[47] that what can be transported is what moves locally, and that something flammable must exist prior to burn-

ing. In physical and generated things, however, not only is what undergoes effects in motion when it undergoes them, but also what causes effects is in motion when it causes them, and in this case too, what is capable of causing motion must pre-exist what moves. This is why he added 'and what is capable of being burned [must exist] before there can be burning'.[48] Clearly, this is not true of every actuality: for it is not [true] of things that are perfect and above what is generated, but of things that are imperfect, in the case of which the actual is always accompanied by what is potential. But such is motion, which is a mixture of both, so that in the case of motion that is transformative, it is true to say that what naturally tends to be moved must exist prior to motion, whether it is moved in a passive or an active way.

> **251a16-28** Therefore, these things must also [have come into being, [after] not existing at a given moment, or be eternal. If, however, each mobile came into being, before the one taken, another change and motion must have occurred, by which what is capable of being moved or causing motion came into being. If, since they always exist, they pre-existed in the absence of motion, this appears absurd to those who attend to it at first glance, but this must be the case even more as they progress further. For if some things are mobile and others movers, sometimes something will be a first mover, and the other thing moved, and sometimes not, but they are at rest,] then this[49] must change previously.

Having assumed that both the movable and the mover must pre-exist motion, and assuming it both from the definition of motion and from self-evidence, he next goes on to require, by division, that 'these things must either come into being at some time, not having existed previously, or else be unengendered'.[50] But if both the movable and the mover came into being, since all generation is through motion, before one caused motion and the other was moved there must have been a motion by which they came into being (for otherwise it is impossible for anything to come into being), so that before the one caused motion and the other was moved, there must be motion. Thus, what was taken was not the principle of motion. But if the movable and the mover did not come into being, but 'in their existence, always pre-existed when motion did not exist',[51] he says that it will immediately appear absurd to anyone who pays attention, that since both the movable and the mover are everlasting, and have not, throughout the infinite time prior to that moment, been active according to the capacities they have, that they
1128,1 should now begin that activity without a cause. For we cannot give any reason why this should occur now, and not earlier. The absurdity will, he says, be even more apparent as we enter more deeply into the argument. For if both movable things and movers existed but were inactive for all previous time, so that there was no motion then, but now

one thing began to cause motion, and the other to be moved, then some transformation with regard to them must have occurred before the cause, through which, although they remained inactive for the infinite time prior to that, one thing now began to cause motion, the other to be moved. But if there was transformation, there was clearly also motion, for all transformation is motion,[52] or accompanied by motion. For if the same things remain around them, they would not previously have remained inactive, but now be moved, beginning from some [point in] time; but some cause, that did not exist previously, necessarily came into being, for the fact that things that were not active throughout infinite previous time, now began to be active with regard to one another. Either they were previously distant from one another and now came closer together, or there was something in between them that prevented [their interaction], which has now been eliminated, or they were not even apt previously, and later became apt. Whichever of these occurs, a first transformation and motion are required, so that of these one can cause motion, and the other be moved, in act. Thus, once again, there will be motion prior to the first motion.

Having said 'for there was some cause of rest',[53] he adds the reason for this, introducing 'for rest is the privation of motion', which had been proved in the preceding discussion.[54] But all privation comes about from some cause, as does possession (*hexis*), and particularly the one that consists in rest. For rest is not the same kind of privation as privations that are contrary to nature, for just as nature is said to be the principle of motion, so it is of rest. If, then, everything that comes into being by nature does so by some causes, it is clear that there must also be some cause of rest. But rest is a privation, in the sense that aptitude by capacity is said to be privation. It, too, is therefore a kind of principle of things in generation, and it is prior in time to possession and to perfection.

251a28-b10 For some things cause motion in only one direction[, others in both of two contrary directions. For instance, fire heats but does not cool, while the knowledge of contraries seems to be one. Yet here, too, there appears to be something similar: for what is cold heats when it somehow turns away and departs, just as a knowledgeable person errs voluntarily when he uses knowledge in reverse. Yet all things that are capable of causing or undergoing effects, or causing or undergoing motion, are not so capable in every case, but only when they are disposed in a certain way and come close to one another. Thus, [there is motion when] they approach one another, and one causes motion and the other is moved, and when they are such that one was motive and the other mobile. If however, they were not always in motion, it is clear that they were not so disposed as to be capable, the one of being moved,

> the other of causing motion, but one of them had to change; this must occur among the relatives. For instance, if what [was] not double is now double, then one if not both of them must change.] Thus, there will be some change prior to the first one.

Having shown that even if one supposes the mover and the moved to be generated, if they are everlasting, but derive their causing motion and being moved from a beginning in time, then prior to the supposed beginning of motion another motion must be assumed, so that there will necessarily always be a change previous to what seemed to be the first change, and thus motion is shown to be everlasting, by the fact that it is not possible to take the first one. Having proved this primarily on the basis of the object in motion, he next proves the same thing from what causes motion. For if what is potentially capable of moving something is to cause motion in act as well, it must first change, with the movable substrate remaining in the same state. And this was immediately obvious, for with the movable object remaining in the same state, if what is capable of causing motion was in the same state as well, why was it sometimes merely potentially capable of causing motion, and at other times in act? But he also proves this by another approach: assuming potentially beforehand that things that naturally cause motion in contrary directions, with the objects in motion remaining the same, must themselves undergo change when causing these contrary motions, he then assumes by division that of things that cause motion, some cause one simple motion, as fire heats and snow cools, and others [cause] contrary motions, as knowledge [moves] the soul toward the true and the false, when the virtuous man deliberately errs in order to deceive someone. And he notes that things that seem to cause motion in only one direction also impose contrary dispositions, as cold accidentally heats by withdrawing. Thus, what causes locomotion also causes rest by withdrawing, that is, by the fact that it is disposed in an altered way. Thus, all things that cause motion can neither always cause motion nor always cause rest, but, for instance, they cause motion when they approach, and cause rest once they have withdrawn. Thus, when they are so disposed that one of them fails to cause motion, and the other fails to be moved, and then later one causes motion and the other is moved, it is clear that a change has come about, involving either both of them, or just one. For this is how all relatives come to be in act from potentiality: either by one of them transforming, or both. But both the mover and the moved are relatives. Therefore, once again, he rightly concludes the argument by adding: 'there will therefore be a change previous to the first'.[55] Again, the same thing must be asked about this [change], by which the mover is moved so that it may be capable of causing motion in act. Again, why was it moved by that motion now, but not previously? For again, what is capable of moving the mover will itself be in motion previously, so that it may move the latter, whereas it did not move it previously. Such, then, are Aristotle's comments in the first of the demonstrations, and they were, I think, said according to such concepts as these.

But since that Grammarian has heaped up a huge pile of garbage, not so much on [Aristotle's proofs] as upon senseless men, let us, with 1130,1 Heracles, summon the Alpheios,[56] and clean out as far as possible the souls that have accepted this trash. Indeed, having set forth Aristotle's text beforehand, and adding to it the entire explanation of Alexander and the paraphrase of Themistius, so that his works would thereby become more voluminous and impress the layman by their dimensions, he then adds his own objections to the argument, of which the first is as follows.[57] He says, 'Aristotle has defined motion as the entelechy of the movable *qua* movable, and he included all motion in common in the definition, but he also supposes that of motions, some are everlasting, while others have a beginning and an end. Why, however, did he assume, as if it followed from the definition of motion, that motion that is not everlasting, but has a beginning of its being, is necessarily pre-existed in time by the things that are to be moved, which have only the potentiality of motion, without the actuality?' He tries to establish that this is not true, as follows: 'Every definition', he says, 'is predicated equally of the entire definiendum', and here too he lengthens his argument with examples. 'Thus', he says, 'the definition of motion, too, similarly applies to beginningless motion, if there is any such, and [motion] that has a beginning. If, then, in the case of [motion] that is not everlasting, it follows from the definition that the object in motion must exist prior in time to motion, then the same will follow in the case of everlasting [motion]. But if this is true, it is necessary either that the definition is not true of everlasting motion as well, although it was given as common, which is absurd; or if it is true in this case as well that the object in motion temporally pre-exists motion, then the substance of the heavens would also temporally pre-exist its circular motion. But since nothing that has something existing prior to itself in time is everlasting (*aidion*), then the everlasting motion that characterises it would not be everlasting, and his argument has been turned around to its contrary. Or else', he says, 'if they wish everlasting motion to exist, then it is not true that it necessarily follows from the definition of motion that the moveable object temporally pre-exists motion, for it does not pre-exist it in the case of everlasting motion'. He then strives to prove that when Aristotle gave the definition of motion, he did not have that concept of what is potential as being separable from actuality, that is, from motion, but [he thought] that motion exists as long as what is potential is inherent in it, and what is potential exists as long as does motion. For once motion has ceased, he claims, when the moving object 1131,1 reaches its goal, what is potential also ceases. And he cites Aristotle's text, which says 'but when a house exists, the buildable no longer exists. Yet it is the buildable that is being built'.[58] And he also calls to witness the exegetes of what is said there, [who say] that Aristotle has taken potentiality as co-present with motion in that passage, not as separate from actuality, as those who wish to 'reason fallaciously' have, he says, now made use of it. Such, then, is the first objection of this man, who strives to prove that it has not been well said that the things capable of motion exist prior in time to motion.

Against these considerations, we must say, in the first place, that the definition of motion is of all motion identically, both that which is everlasting and that which has a beginning and end; for all motion is the actuality of things that, after being active, still have the potential for being active. This is why motion is said to be incomplete activity, because it always co-exists with what is potential. Second, that [the statement that] 'what is capable of being moved pre-exists motion' is also true in the case of all things in motion, but in the case of those that have a beginning and end of their motion, what is capable of being moved pre-exists motion in such a way that it exists even before motion exists. In the case of those that are moved by everlasting motion, however, there is always motion, but it is, of course, different at different times. But what is potential with regard to each motion always pre-exists motion. For when the sun is moving [across] Aries, it has the potentiality of traversing Taurus as well, after Aries, and when it is traversing Aries, it clearly pre-exists, together with what is potential, its motion in Taurus'. This being the case, see how this fellow is either addressing the others as if they were senseless, or fails to sense that he is deceiving himself by fallacious reasoning. For if, he says, the heavens temporally pre-exist circular motion, such motion could not be everlasting. But he reasoned fallaciously, by taking the everlasting circular motion to subsist all at once as a whole in its infinity, and thus supposing that the heavens pre-exist it in time, and he thinks he can conclude that the circular motion is not everlasting, since it has something pre-existing it in time. But what should have been supposed was not everlasting motion *qua* subsisting all at once and infinite, but as
1132,1 having its being everlasting in becoming *ad infinitum*. For this is the nature of motion, *qua* subsisting all at once, but having its being in becoming. Thus, the motion that always exists, but is different at different times, is necessarily always pre-existed by the things that are capable of being moved. It is therefore in vain that this fellow boasts of having turned the argument around to its contrary, and it is also in vain that he says that Aristotle reasons fallaciously, as do the exegetes of Aristotle, when they say that it follows from the definition of motion that what is capable of being moved exists prior to motion. For although Aristotle said that what is potential exists together with motion even when motion has ceased, once the object in motion reaches its goal and the house comes into being, what is buildable, that is, what is potential, ceases. But he did not also say that if motion does not yet exist, what is potential does not exist

either, for this would have contradicted the argument that what is capable of being moved pre-exists motion. This fellow, however, who understands nothing of what has been said, strives to prove that Aristotle and his exegetes say that what is potential and motion both co-exist and cease to exist together, not noticing that even if they co-exist and cease to exist together, there is nothing to prevent them from being preceded. For this most obviously takes place in the case of motions that have a beginning and end: what is capable of being moved by each motion, which is at rest until that point, is moved slightly later. How, then, could the man who takes pride in writing down these utterances by Aristotle in which he says that what is potential co-exists and ceases together with motion, fail to understand those in which he says that what is potential and what is in act are neither in the same respect nor simultaneous, when he says in the third book: 'since some things are the same in potential and in actuality, but not at the same time or not in the same respect, but, for instance, as hot in entelechy and cold potentially'?[59] And again, he is the one who says 'it is possible for each thing to be sometimes active, other times not',[60] that is, of the things that are potential. Yet he also cites Themistius' paraphrase of this text, which says[61] 'for it is possible for each thing to sometimes be active according to what it is capable of, and sometimes not yet, as for instance the buildable that is potential is sometimes active, like the buildable when it is being built, and sometimes not, for instance if 1133,1 stones and mud were simply lying there, not yet being set in motion by the craftsman', but did not notice that these arguments state that what is capable of being moved pre-exists motion.

Such, then, is this man's first fall, which, I believe, displays a great deal of thoughtlessness in the eyes of an upright judge, so that I may omit the large amount of intermediary material, which is affirmed with a great deal of insolence. But perhaps nothing prevents me from mentioning at least one of these, so that his condition may be observed. 'Every definition,' he says, 'is predicated with equal validity of the definiendum, as the definition of substance is not understood differently in the case of everlasting substances, and in the case of those that are in generation and corruption'. Yet who has come to learning so late in life as to be unaware that in accordance with the procession from the one,[62] generated substance proceeds from everlasting substance, and that even if there is some common definition of substance, it too will apply to the various substances in the sense that it proceeds from one thing, but it will not, as this fellow says, apply to various substances with equal validity? But it seems that having heard in introductory lectures that genera are predicated equally of their species,[63] he considered that this also applied to definitions.

He mounts a second assault on what Aristotle says,[64] which is as follows: 'and in the case of motions', he says, 'that are agreed to have a beginning and limit to their being, if some of the things moved in this way were shown to have the motion that accrues to them by nature immediately following as soon as they are brought into being, I think it would become clear to everyone that the definition will <not>[65] apply to their motion either. For being movable and capable of moving, they move in such a way that the capacity still remains, and realities that exist merely potentially do not pre-exist their motion'.

Having proposed the contest in these terms, he seeks to prove that some generated things move immediately by their natural motion as soon as they come into being. And he says that if fire takes form below, it no sooner becomes fire than it receives its locomotion upward, as coincident with its substance; likewise, the water that is formed in the clouds no sooner takes form than it is transported downwards, unless something impedes it. Then, as if he were solving the objection that says that wood, which is potentially fire, pre-exists fire's upward motion, when fire is generated from wood, he says 'it was not the wood that had the capacity of upward locomotion. For the change of wood to fire is generation, not motion; but the abovementioned definition was not of generation, but of motion (for [Aristotle] says that it is the entelechy of the movable *qua* movable). Therefore, if locomotion upwards is motion, not generation, but the change of wood into fire is generation, not motion, even if generation does not take place without motion, the upward motion of fire was not the entelechy of the wood's capacity. For things that move according to each species of motion do so while remaining in their own substance and without perishing, but wood does not remain wood when it is transported upwards; but once it has perished and transformed into fire, it is not [wood] that is moved – for how could what has perished and no longer exists move? – but [what moves is] the fire that has come about from it'. Not content with these, however, he adds other arguments, in an attempt to prove that it is not wood that has the capacity to be transported upwards, but fire. He thinks that he thereby proves that it is not true that what is capable of moving pre-exists motion, [a proposition] on the basis of which the perpetuity of motion was proved. 'Wood', he says, 'pre-exists the upward motion of fire, since it also pre-exists the fire that is generated from it. But fire is no sooner fire than it moves upward'. This, then, is why he persists, having proved that what pre-exists is not what is capable of motion, such as wood, but that the object in motion itself, such as the fire itself that co-exists with the motion, is also capable of motion. 'For if,' he says, 'someone were to say that locomotion upward is the entelechy of wood, first of all, contraries will have the same actuality and motion, for wood is heavy, but fire is light. If, then, motion upward is the actuality both of fire and of the wood that has transformed into it, contraries will have the same actuality and motion, and not with regard to something else, but insofar as they participate in contraries. But lightness and heaviness are contraries. Not only this, but also each of the contrary motions will be the entelechy and actuality of the same thing, since the entelechy of wood will be both downward locomotion, and the upward [locomotion] by which the fire that is generated from it is transported. 'If, then', he says, 'wood, as long as it is wood, does not tend to move naturally by upward locomotion, then it will not have the capacity of upward locomotion, nor will fire's upward motion be its actuality and perfection'. He then adds a third argument with regard to the same point, which goes like this:[66] if all the elements, and the bodies made up of them, are able to transform into one another, either directly or through the intermediary of others (for water might transform into fire through the intermediary of air), the result will be that every motion is according to nature for each one: both each of the light ones and each of the heavy ones. And a fourth one:[67] 'Since', he says, 'growth is the entelechy of what can grow, insofar as it is such, as, for instance, of flesh; and bread and wine, when transformed within an 1135,1 animal, can become flesh, then growth would be the actuality and entelechy of bread and wine. But this is impossible, for what grows is what is nourished, what is nourished are animate beings, and food nourishes but is not nourished. Therefore, the growth of flesh is not the entelechy of food, which, when transformed, became flesh. Thus, neither is the upward motion of fire the actuality and motion of wood, but what tends to move proximately by this motion', he says, 'by its own nature alone, and does not perish, this is what we call potentially movable by that motion. For example, it is man that is capable of walking, not the elements. And we say that motion is the entelechy of the proximate capacity. If, then', he says, 'fire, not wood, is what is potentially movable, then no sooner does it become fire than it is transported upward, if no force impedes it, and the same holds true for water in clouds; then it is not true that in the case of motions that are not everlasting, what is potentially movable temporally pre-exists motion in act'.

In these comments, we must first attend to the goal of what is said: it was contrived exclusively for purposes of counter-argument, with no concern for the truth. For Aristotle proposes to prove that there is no first motion, but that before whichever one is taken there is another one, whence it follows that motion is everlasting, since one is always taken before another *ad infinitum*, and he proves this by assuming beforehand that what is capable of moving pre-exists motion. If it is true that there is no first motion, but false that what is capable of moving pre-exists motion, then the doctrine that motion is everlasting is not refuted, but at most the mode of demonstration will appear unconvincing. This fellow, then, having proposed to eliminate the [affirmation that] what is potentially in motion pre-exists motion, took the motions of fire and water, about which he agrees that their generations from other things pre-exist them, so that there is always a motion prior to motion.

Why do I say these things, when this fellow is obviously blinded by his zeal for contradicting? 'For so far', he says,[68] 'we are not investigating here whether what comes about does so out of what exists or out of what does not exist, nor whether there is or is not some first motion, but that the axiom, actually taken from the definition of motion,[69] is not true'. Yet if there is no first motion, even if the axiom is not true, the doctrine concerning everlasting motion has not been damaged at all. I therefore went into the matter at some length, with a view to exposing this man's goal. But that what this fellow says does not shake the 1136,1 axiom, and that he clearly has not understood what is said, you might learn in this way. What is capable of moving is twofold, and since he was unaware of this, this fellow confused both himself and those who are unaware of his condition. For we say that what can move is what has the perfect capacity that projects actuality, which is accompanied by actuality (thus, fire is capable of moving upward, and water downward, with actuality always accompanying this capacity, and proceeding from it). But we also call 'capable of moving' what is potentially movable, but does not yet have the perfect capacity that produces actuality, but tends by nature to have it. Thus, we say that what is potentially in motion is what naturally tends to be in motion, but does not yet have at the ready the perfect capacity that produces actuality. Therefore, Aristotle said that what is capable of moving pre-exists motion, in the sense of what tends [to move] but is not yet in motion, or is not moved in that part of motion (for we say that the sun is potentially moving in Taurus when it moves [across] Aries). Since, in the case of these things as well, [Aristotle] said that what is capable of moving pre-exists motion, this fellow thinks he refutes Aristotle's argument by proving, in the case of fire and water, that motion co-exists with what is capable of moving by virtue of the perfect capacity that produces actuality. In his ignorance of this difference, then, it is in vain that he declaims everything that follows, striving to prove that what has the capacity of motion upwards is fire, which is already in motion upwards, not wood, which tends to become fire. He fails to conceive that fire has one [capacity], viz. the perfect one, and wood another one, viz. the [capacity] that is imperfect and is said in the sense of what exists potentially, by which it, too, once it has become fire, tends to be transported upwards. In general, if wood is potentially fire, and motion upwards is naturally united in act with fire, how could it not be the case that just as being fire belongs to wood potentially, so moving upwards also belongs to it potentially? But, he says,[70] 'the change of wood to fire is not motion, but generation, but the aforementioned definition is not of generation, but of motion, for he [Aristotle] says this is the entelechy of the movable *qua* movable'. Thus, he would say, Aristotle said the wood pre-exists generation, but not motion. Is this fellow, then, up to the task of opposing Aristotle, he who does not even know what Aristotle has said? For instance, he [Philoponus] says that the definition mentioning 'entelechy of the movable *qua* movable' is of motion, not generation, unaware that Aristotle gave this definition before he distinguished generation and destruction from motion properly so called, for he did this in Book five.[71] He is even more ignorant of the fact 1137,1 that [Aristotle] clearly gave the definition of motion as including generation and destruction as well. And somewhat earlier[72] (and in even clearer terms) [he says]: 'having distinguished in each genus between what is in entelechy, and what is potential, I say that motion is the actuality of what exists potentially, as such, for instance alteration [is the actuality] of the alterable *qua* alterable, increase and decrease of what is liable to increase and its contrary, what is liable to decrease (for there is no common name for increase and decrease), generation and destruction of what is generable and destructible, and locomotion of what is transportable'. Why, then, does this fellow say that the change of wood into fire is generation and not motion? So that [he can say that] what is generable is prior to generation, but what is movable is not prior to motion. But that generation is also a motion, in accordance with what is common in motion, and has been placed under the definition of common motion, is clear at the outset, as is [the fact] that the motion of wood transforming into fire is a generation and a change, viz. into fire, and that wood pre-exists [this change] proximately, with regard to which it existed potentially, but [it does not pre-exist] the upward locomotion of the fire. Nor is the latter the entelechy and motion of the wood, but the change into fire is. This fellow, however, as if Aristotle said that what is capable of motion pre-exists motion, not the [motion] of which it was said to be capable, but the motion of what it transforms into, for instance that wood, which is capable of transforming into fire, pre-exists not the change into fire, but the upward motion of fire, this fellow attempts to rectify the argument by many proofs. Clearly, everything said by someone who objects irrelevantly is superfluous. However, so that I might reveal this man's condition in these matters also, I will devote some discussion to them as well.

'For if', he says,[73] 'someone were to say that the entelechy of wood is locomotion upwards, contraries would have the same actuality, and each of the contrary motions will be the entelechy and actuality of the same thing', and we cited the passage in which he shows these things shortly above. But what, my good sir, is absurd if wood has motion upwards potentially, in that it tends to become fire, while fire has it in 1138,1 act? Again, of wood in act [the capacity] is locomotion downward, but potentially it is locomotion upward, since it is potentially fire.[74] But these things can also be divided into what is *per se* and what is by accident. Or is it not true to say that water cools in act and *per se*, but it heats potentially and accidentally, because water tends to transform at a certain moment into fire? And he should have listened to Aristotle, when he says,[75] in that discussion of motion: 'since some things are the same potentially and in entelechy, not at the same time or in the same respect, but, for instance, hot in entelechy, and cold potentially', it is cold, and potentially downward-tending. But it is clear that wood, as long as it is wood, does not have the capacity that produces upward locomotion, but it is potentially and by aptitude that wood is fire, and is transported upward.

The third [argument], adduced on the basis of the same ignorance, is similarly easy to solve. For all motion, both upward and downward, is appropriate to every element, both light and heavy, and according to nature, if you will, except that 'according to nature' [is to be understood] in the sense that it is sometimes potential, sometimes in act. After all, locomotion upwards is appropriate to water potentially, when it can become fire.

The same is to be said in response to the fourth one as well, viz. that bread, since it is potentially flesh, could be said to be increased potentially. For although the proximate aptitude for bread is that of the change into flesh, as it transforms into flesh it tends to be increased in act.[76] But since this fine fellow declares, as if laying down the law, that what we call potentially movable by a given motion is what tends proximately to move by that motion while remaining in its own nature and not perishing, as what is capable of walking is man, not the elements, we must ask him, in the first place, why did Aristotle say 'hot in entelechy, but cold potentially'? Was it not in the sense that what is potentially cold tends to transform into what is cold? Next, [he must be asked] whether he himself[77] never said that bread is potentially flesh, or sperm potentially an animal, although sperm is turned into a man without remaining in its own nature. Here, too, however, the same thing has happened to this fellow. Ignorant of the difference between perfect capacity, on the one hand, and imperfect potential and the intermediate dispositions, on the other,[78] he attaches those of perfect capacity to imperfect potential. For what is active without change acts by the perfect capacity that projects actuality, while imperfect potential comes to be in act when what is capable is transformed: when the generable is in the process of generation, when what is liable to increase is being increased, and when what is alterable is being altered. But there is also an intermediary disposition, in which what is potential somehow coincides with perfect capacity. It is seen in things that are in local motion, since things in local motion are more perfect than the others.[79] We say, for instance, that fire is movable upwards, with perfect capacity always accompanying the imperfect capacity that approaches nature, since local change takes place without change in the moving object. Yet what changes as it comes into being possesses what is potential (*to dunamei*) in a less perfect state, and the very capacity (*tên dunamin*) by which it changes is less perfect than [the capacity] that causes local motion. Such being the differences between what is potential (*tôi dunamei*) and what is in capacity (*têi dunamei*), [Philoponus] himself, in the case of things in local motion, considers that everything that is potential, remains in its own nature, and does not perish, is said to be potentially movable.[80] But he should have taken

into consideration the potential in the case of generation as well, which changes into a state of actuality by virtue of a change in substance, as [the potential] in the case of increase [takes place] by virtue of [change] in quantity, and in the case of alteration, by virtue of [change] in quality. And in the case of wood transforming into fire, it would have been better to see motion by generation, and [to see] how the wood that is capable of moving by this motion pre-exists such motion both in nature and in time, whereas in the case of fire, [it would have been better to see] local [motion], and how it, too, pre-exists its upward-tending actuality in substance; for once it becomes fire, it must move upwards, even if it moves immediately as soon as it becomes fire. And if someone were to say that wood also pre-exists upward transportation, as it pre-exists its transformation into fire, he would not be wrong. Thus, in the case of the three other motions – that is, generation, alteration, and increase – even this fellow would, I think, admit that what is capable of moving exists prior to motion, as for instance sperm, and what augments in quantity, and what transforms in quality. And in the case of local change, in which this fellow considers he can refute [Aristotle's] argument most of all, since he too agrees that what is potential is inherent in the moving object (but everywhere what is potential pre-exists what is in actuality, even if capacity co-exists with actuality), it is clear that what is capable (*to dunamenon*) pre-exists what is in actuality. After all, fire too, although it moves upward as soon as it becomes fire, possesses in its change what is potential – for instance coming to be present in air – pre-existing what is in actuality, 1140,1
and what is potential in the fiery sphere pre-exists what is in act.

For all these reasons, the Aristotelian argument that says that realities capable of moving must necessarily pre-exist in the case of each motion is completely irrefutable. It is thus in vain that this fellow, boasting at the end of his discourse, says[81] 'if it has been proved false that what is capable of moving pre-exists motion, then what was established on the basis of this at the same time', as the Grammarian said, 'viz., that motion is beginningless, has been refuted along with it'. But I think it would be more just to say that if this [proposition] has been demonstrated to be completely true, then what is proved on its basis would be true, not troubled at all by the clamour of jaybirds.[82]

This fellow, however, having spat on Aristotle's principles of demonstration, wants to introduce to us his own approach,[83] saying 'only thus could motion be proved to be beginningless: if that famous [axiom] of the natural philosophers, as they say, were true, viz. that "nothing comes into being from what does not exist in any way", but everything possesses its generation from something pre-existent'. And he reproaches Aristotle for not taking up this axiom in order to conclude that motion is beginningless, this man who is ignorant of how far he is from Aristotle's sharp-wittedness, and who is also ignorant of the fact that it is proper to Aristotle's science always to take the proximate principles

of demonstrations, not the common ones. For in the first book of the *Physics*, when investigating the principles of generation, he made mention of a certain puzzle that sought to eliminate generation, from the fact that what comes into being must do so either from what exists or from what does not exist. But both, he says,[84] are impossible: 'for neither can what exists come into being (for it already exists), nor could anything come into being out of what does not exist (for something must act as the substrate)'. And he solves the puzzle by saying that nothing comes into being out of what does not exist in the absolute sense,[85] but it does come into being out of what is accidentally non-existent, for it does so out of privation, which does not exist by itself (*kath' hautên*), but in what has been deprived. By this argument it is proved both that there is generation, and that the things that come into being are everlasting and inexhaustible, since everything that comes into being is pre-existed by something else, which is also generated, and after what perishes something is left behind, which is also destructible. Here, however, since he intended to prove not that things in motion are everlasting, but that motion is, he argued from the very definition of
1141,1 motion; a definition that indicates that what is capable of moving always pre-exists motion, as substance everywhere pre-exists activity. Could anyone in his right mind wish to abandon Aristotle's approach, and follow this fellow's reckless fantasies?

Since, however, in the belief that it is easy to demonstrate that the argument that says 'nothing comes into being from what does not exist anywhere in any way' is false, he contends that motion could only be everlasting if [the statement that] nothing comes into being from what does not exist anywhere in any way were demonstrated to be true, let us examine this prattle as well, which he said he set forth completely in the ninth and eleventh [books] of his *Against Proclus*,[86] impressing us with the number of books he has written against illustrious men. He now sets them forth at some length, saying:[87] 'first, even if nature produces what it fashions out of existent things, by virtue of the fact that it has both its substance and its activity in a substrate, without which it is not capable either of being or of acting, it is not necessary for God, whose substance and activity are transcendent of all beings, to create out of existent things. For in that case, he would be no better than nature, whereas God produces not only the forms of the things that are fashioned directly by him, but it is believed that he brings forth and fashions matter itself; for only what is first is ungenerated and uncaused. If, then, God gives existence to matter as well, but matter does not require another matter in order to exist, for it is the first substrate of all natural things, then it is not the case that everything that comes into being does so out of something that exists. For whether matter comes into being from God always or at a given moment, it will certainly have no need of another matter, since it itself is the first substrate of bodies. If what comes into being by nature does so out of existent things, therefore, it is not necessary that the things that come into being directly by God do so out of existent things, since nature needs both some time and [the process of] generation in order to fashion each natural thing, while God gives existence to what comes into being directly by him timelessly and without generation, that is, without forming and shaping individuals. For it is enough for him to will, in order to bring about the substantification (*ousiôsis*) of things.'

He then reminds us of another demonstration, stated in Book 11 of 1142,1 the *Against Proclus*,[88] by which he proved, as he thinks, that both what is generated by nature and what is generated by skill comes into existence from not-being, and perishes into not-being. 'Since', he says,[89] 'the things that are generated in this way are compounds of matter and form, and since prime matter is neither generated by nature nor does it perish, neither does the compound of matter and form in its entirety (for otherwise the result would be that matter is generated and perishes), it remains that there is generation and destruction only of the form. If, then, colours, figures, and the irrational and natural faculties of the soul, and in short enmattered forms in general, have their existence in a part of time (I mean the indivisible forms in each natural reality, for it is of these that there is generation and destruction), if, then, these things are transformed, in the destruction of bodies, into what does not exist in any way, it is clear that they did not exist previously either, and derived their existence out of what does not exist'. After this, he goes on to say: 'But perhaps this will be of no profit to us with regard to the present subject. For even if enmattered forms have been brought into being out of what does not exist, no natural generation comes into existence without motion. For generation is necessarily accompanied by motion, so that if generation always exists, motion has also always existed. Yet whence [does he deduce] that generation has always existed? For whoever says that generation has always existed says nothing other than that motion has always existed. Thus, he begs the question, and takes for granted what was under investigation'. Then, he next adds – incongruously, as it seems to me – 'for neither are the things fashioned directly by God generated out of some previous substrate, nor through generation or temporal extension. For God brought matter itself and time into existence simultaneously with the universe, so that motion did not exist prior to the world. It is clear from this, then, that unless some argument proved the world to be beginningless and endless, it is impossible to demonstrate that motion is everlasting. Since all the arguments by which they establish that the world is everlasting have been refuted, what follows from them is necessarily refuted at the same time'.

But if I have had the patience to copy out word for word so many passages, and of such a kind, let none of the more clever convict me of ill-spent leisure, for I too am aware that the incongruity of these arguments is obvious to those who have been educated in any way. Yet

since this fellow, adapting himself to uneducated listeners, also writes
1143,1 at great length, and draws conclusions, as if they were demonstrated,
about things that have not even reached his understanding, I think it
is humane to indicate the man's condition to them, too, insofar as is
possible. Let us first note whence this lengthy argument began. He said
at the outset that motion could only be demonstrated to be beginning-
5 less if the famous axiom of the natural philosophers, as they say, that
'nothing comes into being out of what does not exist anywhere in any
way' were true; and he maintained that if this is so, it is impossible for
motion not to exist always; and shortly afterwards he boasted that it
was enough for Aristotle to assume this axiom alone, in order to
conclude syllogistically that motion is beginningless. Hear, then, what
10 this wise man, having forgotten these things, added in what follows.
For having proved, as he believes, that generation comes out of what
does not exist, which was his goal insofar as it destroyed the [proposi-
tion that] it comes out of existent things, from which alone he said it
follows that motion is everlasting, he goes on to say: 'but perhaps this
will be of no profit to us with regard to the present subject', and so on,
15 lest I should stir up the same dust once again. But if it is of no benefit
to us with regard to the subject, my fine sir, why did you say that motion
could only be demonstrated to be beginningless if what is now said to
be of no profit for the present subject were true? And why did you say
that 'it was enough for Aristotle to assume this axiom alone', when it
was clearly of no benefit?

20 Let us see, then, how he overturns the famous axiom of the natural
philosophers, proving, on his part, that what is generated comes into
being out of what does not exist anywhere in any way, and perishes into
what does not exist anywhere in any way. And before this, let us note,
if you will, whether this man has understood in general the sense in
which what is generated has been said to come into being out of what
exists or out of what does not exist. For he obviously understands it in
25 the sense that the same form must pre-exist in order for what is
generated to come into being out of it: for instance, matter out of matter,
and individual forms out of similar forms. For he says: 'if, then, God
gives existence to matter as well, but matter does not require another
matter to exist, then it is not the case that everything that is generated
comes into being out of what exists'. And about individual forms, of
which he affirms generation and destruction take place,[90] [he says]: 'If,
1144,1 then, all these things transform, on the occasion of the destruction of
bodies, into what does not exist at all, it is clear that they also did not
exist at all previously, and derived their existence from what does not
exist'. For if the reason why the originated out of what does not exist is
that they did not exist previously, it is clear that if they originated out
of what does exist, they would have existed previously. Nor did he listen
5 to Aristotle in the first book of the *Physics*, when he set forth the puzzle
of the natural philosophers, and said:[91] 'for neither can what exists be

generated (for it already exists), nor could anything be generated out of what does not exist (for something must act as the substrate)', for he did not consider it possible for it to act as substrate for itself. But is [Philoponus] ignorant of the sense in which the things that are generated come to be out of what exists, or out of what does not exist, is said among natural philosophers, but knows what generation is, or is he obviously even more ignorant of the latter point? After all, by 'being generated' Aristotle and the other natural philosophers mean not what comes into existence from any kind of cause, but what has been assigned its passage to being in a part of time.[92] This is why they said it was not possible either for what exists to be generated out of what exists (for what exists is pre-existent), or from what does not exist (for some substrate is necessary), both of which apply to what is generated at some time. For how could what is everlasting and always existent originate out of being or out of not-being? But Aristotle, although he says that the cause of the heavens and their everlasting motion is god, nevertheless proves it to be ungenerated and imperishable. However, when solving the puzzle of the natural philosophers, he says that what is generated comes into being out of its own privation. But how could there be a privation of what is everlasting? And why should I mention the others, when this fellow himself says that generation and destruction are of individuals, which have their existence in a part of time? But the man who says these things also says that matter, which is everlasting, is generated, and in its case he tries to prove that what is generated comes into being not out of what exists, but out of what does not exist. And he says that not only matter, but everything else, except the First, is generated, and he clearly declares that only the First is ungenerated and uncaused. See, then, how having previously called 'generated' individuals, which have their existence in a part of time, here he says that what is generated are the things that depend on a cause, since saying that matter is also everlasting, he says it is generated out of what does not exist, and if he says that only the First is ungenerated and without a cause, joining 'without a cause' to 'ungenerated', and not even respecting those who share his views, he says, namely, that what comes after the First is also generated and created.[93] For he too says that what is generated is fashioned, but not what is begotten. Nor is it when discussing different subjects that he makes these Euripous-like[94] twists and turns, but in one chain of utterances. But what kind of a man 1145,1
should we consider him to be, he who, without even knowing what he's talking about, prides himself on contradicting the leaders of the philosophers, in the belief that such gigantic audacity[95] will go unpunished?

Let us see, however, how he demonstrates that things that are generated come into being not out of what exists, but out of what does not exist, and behold, here again, the self-contradictory character of his arguments.

At the beginning of his demonstration, he agrees that nature creates what is generated out of what exists, and he goes on to add the reason, viz. that it has both its substance and its activity in a substrate, and he says that God differs from nature in that the latter creates what is generated out of what exists, while God does so out of what does not exist. Further on, he reminds us of the demonstration, of which he is proud, that proves that both nature and art create things out of what does not exist,[96] as he said God also creates. However, having previously said that God differs from nature in that the latter creates things out of what exists, and the former out of what does not exist, he said in disgust: 'for if God, too, creates things out of what exists, God will be no better than nature'. Yet since he concedes that matter is everlasting, but says that it is generated by God out of what does not exist, along with the forms that are produced directly by him, these also being clearly everlasting, since the things that are generated and perish are generated by nature according to him as well, he obviously says that everlasting things – which Aristotle demonstrates are completely ungenerated – are generated out of what does not exist. And he has clearly missed the point, for whereas Aristotle calls 'generated' the things that have their subsistence in a part of time,[97] and demonstrates that they receive their generation from what exists, this fellow says that everlasting things – which Aristotle considers are not even generated – are generated out of what does not exist, but he says that they are produced by God in another way, for he considers that what is generated must assuredly perish as well.[98] Even if it is true, then, that God, in whatever way, brings everlasting things into being out of what does not exist, this is by no means an obstacle to Aristotle's argument. For in general, if, as he [Philoponus] says, God gives existence timelessly and without [process of] generation to what is generated directly by him, it is clear that these are not what has been allotted being
1146,1 through [a process of] generation. 'But colours', he says,[99] 'and figures, and, in general, enmattered and individual forms, of which there is generation and perishing, transform into what does not exist in any way. It is therefore clear that they did not exist previously either, and derived their existence out of what does not exist'. Yet what kind of colour, or what kind of figure that has perished, did not perish in such a way that it changes into another colour and figure? For if the body endures throughout its changes, as this fellow concedes, considering it to be prime matter, but every natural body is accompanied by colour, then it is clear that a white body will change [into something] having another colour. The same holds true for figure, for it is impossible for a finite body not to have a figure along with its colour. Yet this man, as I said before, seems to me to be completely ignorant of how what is generated is said to be so out of what exists, and to perish into what exists. For he obviously thinks it is said that [things are generated] out of what is identical in form, and into the same existent thing. He

manifests such a concept when he says that matter will not require another matter, and that forms, which previously do not exist at all, derived their existence from what does not exist, although Aristotle has demonstrated the same things thousands of times: that generation comes from contraries, and perishing [takes place] into contraries. It is amazing, however, how although he concedes that the things produced directly by the first God, as he considers the world to be, are not brought to existence either through [a process of] generation or through temporal extension, he spends so many words and instituted such an impious goal, striving to prove that what is directly produced by the first God has a temporal beginning and end. 'Since, however', he says,[100] 'motion did not pre-exist the world, unless one proves the world is everlasting, it is impossible to demonstrate that motion is everlasting'. Thus, having refuted, as he believes, the arguments establishing that the world is everlasting, he has refuted along with them [the proposition that] motion is everlasting. But that this fellow obviously could not understand what was proven by Aristotle concerning the perpetuity of the world, nor does he shake any of the things demonstrated, I believe I indicated sufficiently in what was written on the first book of the *De Caelo*. Now, too, the fact that he said at the beginning of the argument that the perpetuity of motion can only be proved if someone were to demonstrate that what is generated is generated out of things that exist, but further on he says this is of no profit for the demonstration of perpetuity. Toward the end, however, he says 'unless someone proves that the world is everlasting, it is impossible to demonstrate that motion is everlasting', unaware that this was rightly the mode of demonstrating his own dull-wittedness. For if the world were proved to 1147,1
be everlasting, who in his right mind would continue to enquire whether or not the motion of the everlasting universe is everlasting? For if the heavens were not moved perpetually, they would not exist, since their nature is eternally moved, nor would sublunar things remain in constant change. Yet it was characteristic of Aristotle's genius to prove both that the world is everlasting, and that motion possesses perpetuity by itself and from its own nature, as he will display with regard to time. This fellow, however, contriving to lengthen his arguments in order thereby to impress the stupid, makes his own dim-wittedness even more manifest. Retreating once again, he says he has demonstrated that some generated things do not temporally pre-exist their own natural motion – which, as we have seen, he did not demonstrate – and he tries to demonstrate the same thing again, and to thoroughly refute Aristotle's argument stating that among relatives, either both or one of the two must move, in order to receive the relation of mutuality.[101] 'For since', says this fellow,[102] 'Aristotle assumes it to be agreed that relatives must first pre-exist, and then receive the mutual relation, so that it is not possible for things to be relative as soon as they exist, he thus concluded that it is absolutely necessary for either both

of them, or one of them to have been moved, in order to receive the relation of mutuality'. And again, citing Aristotle's text and the commentary of Alexander, he undertakes to demonstrate the same thing, that fire, if generated below, and water, if it comes into existence above, have their motion immediately, as soon as they are generated, and the motive nature of the substrate does not pre-exist [the motion] in time. 'If, then', he says, 'the natural capacity within is what causes motion, and the body that acts as substrate to it is what is movable, and the movable and what causes motion are relatives (*pros ti*), and these must, as Aristotle says, pre-exist their mutual relation, then fire will not immediately move upward as soon as it is fire, if it happens to have been generated below, which is evidently contrary to the phenomena. And in the case of the parts of animals', he says, 'they assume the relation of right and left as soon as they have been thoroughly formed, and it is not the case that limbs are first generated, and then assume their relation'. Yet why should I recapitulate the nonsense of this fellow, who has, I believe, understood nothing of what Aristotle has said, and therefore replies to him irrelevantly? For Aristotle, having said that the realities
1148,1 capable of moving by each motion pre-exist the motion, then goes on to question the argument by means of division, saying:[103] 'Moreover, these things also must either have been generated at some moment, having not existed previously, or they must be everlasting. But if they were generated, another change and motion must have been generated before the one taken into consideration. Yet if they existed and always pre-existed when motion did not exist, it is clear that they were not so disposed as to be capable, one of being moved and the other of causing motion, but one of them was changing (...). For this must occur in relatives, for instance if something was not double but is now double, one or another of them, if not both, must change'. You can see that following the sequence of the division, having supposed that what is capable of causing motion or being moved is either generated at some moment or is always existent, he states what follows from [the hypothesis that it] is generated, and places all the others under the hypothesis that says it is always existent. Thus, <in the case>[104] of the relatives as well, even if substance really always pre-exists relation, this is not what Aristotle says here. Instead, [he says that] since substrates that are capable of moving and being moved always exist, but are at rest, it is clear that they were not disposed such that they were capable, one of being moved in act, and the other of causing motion – for in that case they would not be at rest –, but that one began to cause motion, and the other began to be moved. If, then, one is to cause motion, and the other is to be moved, either both or one or the other must change, taking on an addition or a diminution, so that one may become double and the other half, or approaching if they had been previously distant. What need was there, then, for you to expend so many arguments, proving that relatives do not pre-exist their relation, and to slander Aristotle

by writing:[105] 'It is therefore not true that relatives, if they do not belong to what always exists, must necessarily pre-exist their relation in time', when Aristotle has clearly said: 'yet if they existed and always pre-existed when motion did not exist', and included the testimony of the relatives in this section of the division? But since now too he whinnies with happiness because he has demonstrated that what Aristotle said, viz. that what is capable of motion pre-exists motion, is false, and he seems to provide another argument in addition to the previous ones, let us not leave this one unexamined, either. 'For if', he says, 'the natural capacity in fire causes motion, and the body that acts 1149,1
as substrate for it is movable, and if one causes motion and the other moves as soon as they become fire, neither does what is capable of causing motion and of moving pre-exist motion in time, nor do relatives pre-exist their mutual relation, if what causes motion and what is moved are relatives'. Against this, it has been said previously that if what is moved is a substance, and motion is activity, and substance everywhere pre-exists generated and perishable activity in time, it is clear that Aristotle spoke rightly. For let fire be generated below and be moved upward: unless it assumed the form of fire below, it would not have [the property] of moving upward. If, however, when it exists as fire, it is not immediately in the middle or upper place, it is clear that the substance of fire temporally pre-exists its locomotion and the motion that will continue to occur, and even if you divide it *ad infinitum*, *qua* continuous, what is capable of motion always temporally pre-exists motion, with time, of course, being divided along with it. Thus, even if a natural body is causing motion and in motion at the same time, the substance of what is causing motion and in motion pre-exists its activity. This has already been said: but he who makes such counter-arguments should also know that the natural capacity inherent within the object in motion is not observed in the proper sense in the sense that it causes motion, but in the sense that it is moved, since it is potential and imperfect. And thus it is said to be the principle of motion in that in which it is present, in that it is a principle in the sense of being moved. For what causes motion properly speaking is perfect and tends to be transcendent of what is moved, as what is in act [tends to be transcendent] of what is potential. It is therefore necessary that what causes motion pre-exist what is moved, if indeed what causes motion brings the object in motion from potential to act. Nor do I say these things on my own account, but Aristotle in the third book of the *Physics* says that it is not the nature in the process of being generated that produces man in the proper sense, for that nature is generated more than it produces, but that 'man in entelechy produces a man from what is potentially a man'.[106] And in this book, he will clearly state[107] with regard to natural bodies, that 'none of these moves itself, but it possesses the principle of motion, not of causing motion or of producing, but of undergoing'. Thus, investigating what it is that causes motion in

the proper sense in the case of objects in local motion as well, he did not identify the nature within the object in motion – because this nature is more moved than it is a mover – but what produces its form. For he said 1150,1 that what produces fire is the efficient cause of motion upwards, so that according to these arguments as well, I believe, the objection based on nature and the underlying body, which this fellow raised, will not at all trouble Aristotle's argument, which states that everything that is capable of causing motion and being moved temporally pre-exists motion. After all, in the case of generation, increase, and alteration, what produces and is generated clearly pre-exists: for instance, as sperm and menses, and stones and wood [pre-exist] the generation of a man and of a house, and bread and flesh [pre-exist] increase, and body and sun [pre-exist] the darkening [of the skin]. I believe that this has been proved to be so in the case of local change as well. In the case of the relatives, it has been proved that he reproached Aristotle in vain for saying that in the case of all the relatives, the relatives pre-exist the relation, and that what causes motion and what is moved also belong to what is relative in this sense.[108]

The difficulties raised on the basis of the parts of animals have also been raised in vain. For even if the relation co-existed in their case, Aristotle's argument is not impeded at all, for he himself would say that substantial existences pre-exist active relations. Yet having put forward an objection to himself near the end,[109] to the effect that even if it were not necessarily, in the case of the relatives, that realities pre-exist their relation, they still arrive at being through motion – for the parts came about when the sperm was set in motion, and the sperm when something else was previously set in motion. Likewise in the case of fire and the other things, so that one cannot identify an initial motion. Having objected in this way, then, he again solves it, saying that 'even if nature creates things from what exists, this does not mean that God does so. For if the world did not always exist, it is clear that God created it from what did not exist, and that, if he produces things in a manner similar to nature, he will be no different from nature; and that he produced the matter of bodies from what did not exist, for it was certainly not out of a substrate'. And these things were said by him before, and at length, and I believe they have been solved, but now too, for those who encounter them at various times and in various places, something must briefly be said. And because we are investigating whether motion is everlasting or not, to say 'for if the world did not always exist, it is clear that God created it out of what did not exist' is, I believe, something that is just too senseless, for Aristotle says that the things produced directly by God, being necessarily everlasting, are not even generated, for the only things that are generated are those that are in a part of time, and succeed one another in perishing.[110] If, then, God produced matter, he produced it not from a substrate, but [as] everlasting, as he did the common form. If, then, in natural generation

something posterior is always generated from something previous (for 1151,1
this fellow concedes this), one must continue *ad infinitum*, postulating one thing before another. For if we postulate something first, which was generated directly by God and not by nature, in the first place it must be everlasting and not coming into being. Next, how could what is generated directly by God be of the same nature as the things that are generated from nature? Perhaps they would even be inferior, since nature produces its products from a substrate alone.[111] 'But', he says,[112] 'as this matter for instance became fire from a pre-existent fire, and that one from another, but it is possible to stop at some point as we ascend to some fire that was not generated when its matter was ignited from another fire, but either by friction or some other cause, but not from fire, so in the generation of one thing from another it is not impossible', he says, 'to see something very similar occurring. Now, then', he says, 'as one thing receives its generation from another, inasmuch as they are generated by nature, but it is nevertheless possible that they have a beginning of their existence and that there should be something first in the case of each species, which did not derive its generation from something else that pre-existed, whether similar or dissimilar. This is especially so in the case of the primary elements themselves, but they came into existence at the same time as the first formation from God. For the things that derive from the elements, even if [they first came into being] then primarily, were, of course, generated out of them and in that way. Therefore, as we said before, unless the world could be proved to possess a temporally beginningless subsistence, it cannot be proved that motion is beginningless'. This, then is what that fellow wrote, in his very words.

Let us look at each of his statements. The first thing he says is that even if fire received its genesis not from fire but from friction, there must pre-exist what it was generated from,[113] such as the air that was subject to friction. Again, the genesis even of the primary fire is out of what exists. Thus, if someone were to maintain the same thing with regard to the others, it would in any case be necessary for something to pre-exist what is said to be the first thing to be generated, viz. that from which it was generated. But to say that the elements, which are now generated by nature, come into being out of each other, but in the beginning they were generated by God, is [characteristic] of a soul that is astonishingly lazy in the search for the truth. For whence do they possess this [property of] being generated out of each other, if not from God? Who in his right mind could conceive of such a change in God, such that not creating previously, in the briefest stretch of time he would become the creator of the elements alone, and then cease from creating 1152,1
again, handing over to nature the genesis of elements out of each other, and of the other things from the elements? It is clear, however, that according to this fine fellow's teaching, God is responsible neither for the world nor for everything in it, but only for the elements, and for

those that are perishable,[114] since they change into one another. And yet, as I said, the things that are generated directly by God, insofar as they are generated by him, who alone is unchangeable, always the same, and identically disposed, are also everlasting. Thus, the celestial things, produced proximately by him, have received a subsistence everlasting in number, and of sublunary things, the common features have received eternity in regard to form, since it was in accordance with form that they were produced proximately by him. What is generated and perishes, however, has been woven together with them from the causes in heaven that are everlasting but in motion. And this is what Plato's Demiurge[115] said to the new gods: 'And you, then, interweaving the mortal with the immortal, make and generate living creatures, and give them food, and make them increase, and receive them again once they perish'. But I think that what this fellow dared to counter-argue against Aristotle's first proof about the eternity in motion has been sufficiently verified. Moving on to the second proof, I will try to clarify it, then I will set forth the objections against it, and I will present the sense in which they have been stated unsoundly.

> **251b10-18** In addition, how can there be before and after [when there is no time, or time when there is no motion? For if time is the number of motion or a particular motion, since time always exists, motion must also be eternal. As far as time is concerned, however, everyone appears to speak with one mind, except for one,[116] in saying that it is ungenerated. It is by means of this that Democritus, at any rate, proves that it is impossible for all things to have been generated,] for time is ungenerated.

He also provides this demonstration, also very evident, of the fact that motion is neither generated nor perishes, but is everlasting, establishing it on the basis of time by means of a hypothetical syllogism such as this: if time is everlasting, then motion is also everlasting. But the antecedent is true; therefore, the consequent is also true. And he will prove the hypothetical syllogism both now and later, from the fact that time is motion, as some have considered, or number, or some affection of motion. That time is everlasting, he tries to prove both from the [arguments] that motion is generated, together with inverting the argument. If, he says, motion is generated after having previously been non-existent, and if it later perishes, no longer existing, it is clear both
1153,1 that what is earlier is before motion, and what is later is after motion. But if what is earlier and what is later exist, then time must also exist before motion and after motion. For this earlier and later, according to which they say that motion, not existing earlier, was generated later, and existing earlier, later does not exist, indicates the order of the parts of time, for these are the parts of time, and time is divided into them.

But if time must exist before motion and after motion, but time is motion or[117] the number of motion, as has been proved, then motion must also exist prior to motion and after motion. Thus, for those who say that motion, not existing previously, exists later, and previously existing, later will not exist, it follows that [they must say] that there once was motion when there was no motion, and there will at some moment be motion when there will not be motion. Yet he omitted to add this; but having assumed that time is either motion[118] or the number of motion, inasmuch as whenever there is time there is motion, and that time is everlasting, he added that 'motion must also be everlasting'. He hinted at the fact that time is everlasting by [saying] 'before' and 'after', for if time is what is before and after, but the before and after are everlasting (for if something were prior to what is before, it too is before, and if it is later than what is after, it too is after), then time too is everlasting. He proves this, first, from the common opinion on this subject of the natural philosophers, all of whom, according to him, think time is everlasting, except Plato, for he says it is generated at the same time as the world. We will soon see in what sense Plato says this. Democritus,[119] for his part, was so convinced that time is everlasting, that when he wanted to prove that not everything is generated, he made use of the [proposition] that time had not been generated as self-evident. Anaxagoras too, when he says 'All things were together', indicates time before the world-constructing separation.[120] After all, [the word] 'were' is temporal, and he says that since [all things] were indistinct for the infinite previous time, Intellect later distinguished them and set them in order. Empedocles, who says that the predominance, now of Love and now of Strife, alternates from perpetuity, clearly thinks that time is everlasting. But if time is everlasting, so is motion.

> **251b17-19** But Plato is the only one to say[121] that it was generated, for he says that it was generated at the same time as the heavens, and that the heavens were generated. 1154,1

It is now necessary to state what I have often said elsewhere, viz. that since 'generated' and 'ungenerated' have many meanings, and Plato and Aristotle use them in different senses, they seem to be contrary to one another, although they are not really opposed. After all, 'generated' means what earlier does not exist, but later exists, and what has its subsistence in a part of time, and this is the meaning in which Aristotle uses 'generated', which he places opposite to 'everlasting' in his division. Another meaning of 'generated' is the one that is opposed in divisions to true being, which is eternal and self-subsistent:[122] it is what has its being in becoming and comes into existence from another cause, not by itself. And 'generated' is said by means of both of these, viz. by the opposition to what is truly existent and simultaneously whole, and

the opposition to what is self-subsistent, even if it is everlasting. And it is according to this meaning that Plato calls the entire sensible and corporeal formation 'generated', for all that is corporeal is dispersed, and can neither give existence to itself, nor be brought together into a simultaneous whole, neither with regard to substance, nor to the being of substance. He clearly opposes at the outset what is generated to what exists, where he says:[123] 'What is that which always exists, having no coming into being, and what is that which is always becoming, but is never existent?'. It is, then, in accordance with this [sense of] 'generated', and not the one stated by Aristotle, that Plato says both the world and time are generated. Certainly, having asked[124] with regard to the world 'whether it always existed, having no beginning to its coming into being, or whether it was generated, starting out from some beginning', he answered:[125] 'it was generated. For it is visible and tangible, and has a body, but all such things are sensible, while sensible things, being graspable by opinion together with sensation, have appeared to be in the process of becoming, and to be generated'. You can see that when giving the reason why the world is generated, he did not say that 'for not existing earlier, it later exists', but that 'it is visible and tangible, and has a body'. Yet he would also have it that time is generated, saying[126] 'time, however, came into being with the heavens'. And the reason for this is not that time exists earlier, but later does not exist, for he says it is everlasting, albeit not in the same sense as eternity, but because it is the image of eternity, and came into existence in accordance with motion. Certainly in Plato, Timaeus says,[127] following the corporiform ensoulment: 'but when the father who had engendered it realised it moved and was alive, having become an image of the
1155,1 everlasting gods, he was filled with wonder, and in his joy he bethought himself to make it even more like the model. Since, then, the latter happens to be an everlasting living being, he undertook to render this universe such as well, as far as was possible. The nature of the living being, then, happened to be everlasting, and it was not possible to attach this [property] completely to what was generated. But he bethought himself to produce some moveable image of eternity, and as he set the heavens in order, at the same time he produced, of eternity remaining in one, an eternal image proceeding according to number, the very one to which we have given the name of time'. Note, then, that the Demiurge, wishing this universe, too, to be as everlasting as possible, granted it, since it was like an image, time as an image of eternity, 'so that', he says,[128] 'the world might be as like as possible to the model. Indeed, the model is existent for all eternity, while the world has come to be, exists, and will exist throughout all time'. If, then, throughout all time the world has both come to be and is and will be, it is clearly because it too is everlasting, but not *qua* eternal. For what is eternal possesses all its substance, its capacity and its activity at once, because eternity has brought 'always' together with 'what exists'.[129] But

time and what is in time, since they possess their being in becoming, unwind[130] their integrality according to motion and coming-into-being. If, then, Plato too says that it is through time that the all participates in the everlastingness of which it is capable, this is clearly because time is everlasting according to him as well, albeit not in the same sense as eternity is. Therefore, let no one, by looking at the literal sense [of the texts] consider that the philosophers oppose one another. For if, taking the same meanings of 'generated' and 'ungenerated', one had said that time and the world are generated, while the other said they are ungenerated, they would have been speaking contrary to one another. As things are, they agree in their concepts; for if Aristotle says that time is the number of motion, while Plato says it is an eternal image proceeding according to number, in what respect do these two [philosophers] differ on this point, when they agree right down to their terminology? And if Aristotle says time is everlasting, while Plato says the world has become everlasting through its participation in time, and if Plato clearly says the world is generated in the sense of being corporeal, while time [is generated] insofar as it has its being in motion and coming-to-be, with which Aristotle agrees in both cases, how can we still consider that Plato's 'generated' and Aristotle's 'ungenerated' contradict each other in their concepts, and not merely in their terminology? 'Why, then', someone might say, 'does Aristotle say that all except one say that time is ungenerated, hinting at the beginning, but later adding nominally that Plato is the only one to say it was generated, and adding the reason why it was generated: for it was generated at the same time as the heavens, but he says the heavens were generated'? The answer is that it was ancient usage to argue against the surface meaning out of consideration for more superficial understandings. Since, then, 'generated' was said more habitually of things that did not exist earlier, but later existed, therefore, arguing against this meaning of the term, Aristotle seems to censure Plato for having said 'generated', 1156,1
but in fact he is censuring not Plato, but those who attached 'generated' in this sense to time and to the world.

251b19-27 If, then, it is impossible for time to either exist or be conceived without the now[, and the now is a kind of intermediary that possesses both beginning and end at the same time – the beginning of the time that will be, and the end of the time that is past – there must always be time. For the extremity of the time taken to be last will be in one of the nows (for nothing can be taken in time outside the now). Thus, since the now is both beginning and end, there must always be time on both sides of it. Yet if there is time, it is evident that there must also be motion,] since time is a kind of affect of motion.

Since time is either motion, or a number, or some affection of motion, or simply something that necessarily co-exists with time, if time is everlasting, then necessarily motion is also everlasting. Having previously proved, then, that time is everlasting from the before and the after and, as he says, from the opinion on this subject of all natural philosophers except one, he demonstrates the same thing by another proof that is even more appropriate to time. The demonstration seems to me to be as follows: time exists and is thought of according to the now, for of time, only the now is in subsistence, for the past no longer exists, while the future does not yet exist. What has come into subsistence according to the now subsists by a mean term, which is the beginning of the future time, and the end of the past one. What has come into subsistence according to such a mean term, so that there is both always something before what is taken, and after what is taken, is such; for every now that is in subsistence is intermediate between times, and the nows they contain. Such a thing, then, is everlasting; for whatever one postulates as the last now is itself also intermediary, and has a time after it, so that there is no last now or time. For if there is time, there is a now, since the limit of time is a now, as the [limit of] a line is a point. Similarly, it will have no beginning; for the now that seems to be in the beginning is itself also intermediary: as it is the beginning of the future, so it is the limit of the past. Time is therefore everlasting, since it has neither beginning nor end. But if time always exists, it is clear that motion will always exist as well, since time is the number or affect of motion, and just as it is not possible to grasp motion without time, neither can time [be grasped] without motion.

What Aristotle has said here too, intellectively and divinely, with a view to demonstrating that motion is everlasting, has received as much clarification as possible from the exegetes. But since the Grammarian[131] has once again tried to argue against them, let us see what kind of a man he will turn out to be when put to the test. For this fellow, having agreed that motion is similarly disposed to time as far as everlastingness, the process of becoming, and destruction are concerned, fights against the everlastingness of time, striving to overturn the arguments that establish it, which are divided into three. The first of them carried out a proof immediately from the before and after. For whoever says that motion is generated says nothing other than that not existing before, it exists later, and whoever says it is perishable says nothing other than that it exists before, but later does not exist. If, then,
1157,1 the before and after are parts of time, there will be time before motion and after motion; and wherever there is time, there is necessarily motion. Thus, before what seems to be the first motion, there will always be motion, and after what seems to be the last one, there will again be motion. Motion is therefore everlasting. In reply to this argument, this fellow assumed beforehand three axioms, as it were, with a view to his astonishing construction. The first is that time has

its existence at three removes from bodies,[132] or rather at four, as he says later, correcting himself; for [he says] body is first, second its kinetic capacity, and third motion, and thus in the case of active motion, time is the number of motion. Second, he says that the intellective substances are transcendent of any body, both in substance and in activity, and he praises Plato, Aristotle, and the other philosophers who have demonstrated this. Third, 'it is agreed', he says, 'with regard to the intellect,[133] that[134] its intuition toward the intelligibles happens to be partless and unextended,[135] and does not require syllogisms for the apprehension of the intelligible, as if it passed from premisses to premisses by discursive thought, but [the intellect] either grasps or fails to grasp [the intelligible]. Once again, however, it has also been conceded about the activity of the intellect that it does not receive the intellection of all beings all at once, for it cannot think of God, and an angel, and the differences and commonalities of souls, and bodies, and time, and eternity, and the rest of beings simultaneously. For this is the proper characteristic only of divine intellection, if I may say so, viz. the partless knowledge of all beings, simultaneously those of the past and of the future. But we are not now discussing the divine intellect', he says, ' but the one that comes after it, whether it be angelic, psychic, or of whatever other kind. But intellect', he says, 'since it proceeds transitionally from thought to thought, does not allow two identical thoughts in the same instant, but it thinks one thing first, another second, and another one third'. Having taken these beforehand as axioms, he adds: 'If the intellect is separate from its relation to the body in both substance and activity, it is clear that it is *a fortiori* bereft of the motion of bodies and of time. If, then, intellect as such is unrelated to time, and does not receive the intellection of all the intelligibles at the same time, but having thought this thing first, for instance, it next proceeds to another thought, and again from that one to another, there is therefore an earlier thought and a later one in the thoughts of intellect, although it is estranged from every temporal relation. It is not the case, therefore, that that which whatever way is thought or said to be before or after in whatever way automatically indicates time as well. Since time, too', he says, 'was generated with the heavens according to Plato,[136] if one were to eliminate bodies, he will eliminate time as well, but he will not eliminate the intellect. If, then, the activity of the intellect is transitional even if time does not exist, it is necessary to think one thing first, and then a second thing. This [property]', he says, 'belongs to it essentially. It is therefore not the case that whatever is said in whatever way to be earlier and later is automatically indicative of time. But 1158,1
if', he says, 'someone were to concede that not even intellective substances are free from the relation to time, certainly he would necessarily still consider the creator of time to be higher and beyond all time. If, then, God directly creates the heavens, he also granted them to be moved in a specific way.[137] But the motion of the celestial [bodies]

is not of equal velocity. If, then, one were to conceive of the wandering stars coming into conjunction beneath the same part of the zodiac,[138] it is clearly not the case that each of them will return to the same point in the same [period of] time. Accordingly, when the moon, having accomplished its circular orbit, has arrived at the same point at which all the planets happened to be at the same time previously, God, who arranged these things, no doubt knew that [the moon] had accomplished its circle, but the rest [of the planets] had not yet done so. Thus, if he knew that one of the stars returns to its starting-point earlier, and another later, then, as far as Aristotle's proofs are concerned, one thing is earlier and another later in the divine thoughts as well, and therefore so is time. For they will not say that he is unaware of these things, of which he is the direct creator. But [time is not present] only in his thoughts, but also in his acts of will:[139] for once the moon has occupied the same point, no doubt God willed it to be so disposed, but he did not will the rest [to be so disposed] yet. And when he willed the sun to remain in place, he did not will this for Mercury, let us say, or any of the others. Thus, even in God's acts of will there is a first act of will, and a second, and a third, and therefore there is time. In addition, God has knowledge of future things.[140] Now, what is future is not universals, but particulars and individuals, such as Socrates, before he is generated, and Plato. Accordingly, if God knows the future, he knows that they have not yet become present, but again when they become present, he knows that they are present. And when they have perished, he knows that they have passed and are no longer among beings, so that the knowledge by which he knew they were going to happen is prior to the one by which he knows that they are no longer among beings. If, then, through the expression 'before and after', time is immediately revealed as subsisting coordinately, we shall not place even the creator of time outside of temporal relation'.

Having said these things, he then[141] tries at length to establish that since we cannot conceive or express what is above time, we make use of temporal terms. For as God thinks what is subject to time timelessly, so we [think] what is above time temporally. If, then, we say 'time did not always exist', or 'it did not exist before it was generated', or make use of some other expression that signifies the concept of time, it is not immediately necessary that time be conceived along with the expressions, but merely the fact that time does not always exist. For that
1159,1 temporal expressions do not necessarily signify time, however they may be said, he says he has demonstrated in Book 5 of the *Against Proclus*; and he asks his readers to select their proofs from there, as the Grammarian says, in order to avoid repeating himself (although that is what he enjoys most).

We have quoted most of these arguments word for word, expecting that many would not believe that someone who thinks and writes this way dared to argue against the leaders of the philosophers.

But let us look at what he says, starting out from what he assumed as axioms. At first, he said that time has an existence at the third or fourth remove from bodies, taking into consideration only the time that measures activity. Yet time[142] measures not only the activity of divisible and generated things, and in general of those things that have departed from true being, but also the extension of being and [the extension] according to qualities.[143] Next, he failed to understand – he, who constantly adduces Plato – that [Plato] says that the world becomes more similar to the model through its participation in time, when he writes:[144] 'but when the father who had engendered it realised it moved and was alive, having become an image of the everlasting gods' and so on, with which I do not need to burden [my readers], since I quoted it shortly above.[145] If, then, it was through time alone that the world participated in everlastingless – which this fellow thinks is suitable only for God – and in greater likeness to the model than when it was merely corporeal and ensouled, how could this fellow consider it to be feeble? If, however, it [sc. time] stands analogously to the world as eternity does toward the world's model (for as [the model], through its participation in eternity, which has a higher rank, is eternal, so the world, since it is everlasting through its participation in time, participates in time as in something higher) [146] (...) but it is not surprising if it has escaped this fellow that his idle chatter has been crushed by the finest of philosophers.

The second of his axioms, which says that intellective substances are transcendent of all[147] body in both substance and activity, was well stated, although this fellow is obviously completely ignorant of intellective substance, as we shall learn from the third [argument]. For the same fellow says in the third one, both that [the intellect's] intuition toward the intelligibles happens to be partless and unextended, and that it does not simultaneously know God, an angel, and the differences and commonalities of the souls, and so on, but, proceeding transitionally from thought to thought, it does not admit two thoughts at the same time in the same respect. For he says this is proper to God, but not to the angelic or the psychic intellect, for the latter thinks one thing first, another second, and another third. How, then, can its intuition be partless and unextended, if it does not think two things at the same time, but one thing first, another second, and another third, making a transition from one to another? Which intellect does he mean by this? It is either the rational soul or its summit, or some other substance above the soul. If it is the rational soul that he calls 'intellect' (for it is to this soul that syllogistic thoughts, and those that make a transition from one thing to another, are proper), then this soul is not unrelated either to bodies or to time: for how could a transition from one thing to another take place atemporally? How could that which does not think simultaneously fail to think one thing in an earlier time, and another in a later time? For let the Grammarian tell us what is meant by

'simultaneous' and 'not simultaneous', unless he postulated 'simultaneous' (*hama*) as identical to 'together' (*homou*), being unaware of the difference between them as well. For I sense that, like the man who set his heart on horns according to the myth, this fellow, having set his heart on arrogance in philosophy, has been robbed of his ears as well,[148] [that is], of the appropriate training in grammar, since he considers that what does not know simultaneously, but makes transitions from one thing to another, thinking one thing earlier and another thing later, does not think according to time. But if he considers intellect to be the summit of the rational soul or some substance above the soul, if it carries out its thinking transitionally, as this fellow considers, saying that except for God, at most, all things know in this way, then that is neither above the soul nor above time. But if its knowledge is all at once and without transition, as is fitting for the substantial intellect, there is no longer any place for the before and after, nor for what passes from one thing to another. Thus, he could not find such a before and after unless it were said according to time. But in striving to demonstrate this, he has revealed his confused conception of intellective substance, for he should either have paid heed to the meanings of before and after divided by Aristotle in the *Categories* and slightly later,[149] or, having divided [them] himself, show which ones concern time, and which do not. In order to articulate that man's confusion, however, let it be said that 'first', which indicates order, could be said in three senses in the present passage: either by position (as we say that the fixed heaven is first from above), or by nature and substance (as what entails destruction along with it, but is not destroyed along with [something else], and implies something but is not implied by it, as genera are to species), or what is previous in time, as the Trojan War is to the Peloponnesian War. Let him therefore tell us, if the before and after in the thoughts of what he calls the intellect is not according to time, according to what else would it be? For there is no position among the intelligibles, unless it were said metaphorically. And even if there were, nothing would prevent the before and after from being conceived simultaneously; for I suppose [the intellect] simultaneously conceives the centre and circumference of a sphere. But not only is the intellect forced to conceive simultaneously of what is before and after by nature, such as cause and effect, but so is the soul, because of their mutual relation. What is before and after by transition, and not simultaneous, necessarily has time and motion co-existing with it. And with regard to the before and after in knowledge, which does not exist simultaneously, I cannot even conceive how it could be said in any other meaning than according to time. Yet
1161,1 this fellow exhales empty arguments without reaching an understanding of them, he who having placed intellect above time, gave it transitional activities. For he says,[150] 'having thought this thing, it next goes on to another thought, not thinking them simultaneously', and having postulated this, he says such an activity is above time, clearly

indicating, to those capable of understanding, that he does not have an articulated concept either of intellective substance or knowledge, or of the nature of time. Therefore, intending to show that there is something prior and posterior that does not exist according to time, he propounded that intellect is above time, and is active according to some other [sense] of prior and posterior. Having assumed the temporal prior and posterior, however, he confused everything with it. For if it passes from one thing to another, it is clearly in motion, and where there is motion, there is time, as he too agreed at the beginning of his counter-arguments, when he postulated time to be the number of motion. Even if, then, not every prior and posterior, in whatever sense they may be said, are immediately indicative of time (for the [meanings] 'according to nature' and 'according to position' are not), at any rate, the prior and posterior in transitional thought – in which [the intellect], having thought one thing earlier, next moves on to another, and thence again to another, whatever what thinks in this way may be – this prior and posterior according to time, and not according to any other meaning of prior and posterior, is motion, or is accompanied by motion. For transition, as I said, is motion; but where there is motion there is time. Thus, all his subsequent arguments, which say that the activity of the intellect or of God are transitional but timeless, have been argued in vain. Also vain is the conceptual elimination of bodies and the co-elimination of time. 'For if', he says,[151] 'one were to eliminate bodies, he will eliminate time as well, since time was generated with the heavens, as Plato says, but he will not eliminate the intellect. If, then', he says, 'the activity of the intellect is transitional even in the absence of time, and it is necessary to think one thing earlier, and another thing second, it is not the case that everything said to be before and after is immediately indicative of time'. Yet he also added here that thinking one thing earlier, and another second, belongs essentially to the intellect. And it is perfectly clear that the intellect thinks what is first and second by nature, by position, and by time – of what could it be unaware, since it is intellect? – but it does not think one thing earlier, and another later transitionally. And this may have been what tripped up the Grammarian. For perhaps he had heard the philosophers say that the intellect, having shown forth the first distinction within itself, both produces realities as distinct, and thinks the first beings in a way that is distinct but without transition,[152] and the second beings *qua* distinct, this fellow, not understanding what was said, considered it was said that it thinks the first things, and the others by passing transitionally from one thought to another. This befell him because of his ignorance of what is rightly believed about the intellect. For if he had heeded Aristotle saying about the intellect 'being activity in its substance',[153] and that 'the intellect (...) is the realities',[154] and if, having heeded this, he had understood what was said, he would not have conceived of the activities of the intellect as transitional. For if the activity of the intellect is

compounded both with its own substance and with realities, it is clear that it would be without transition. But it may be, as I said, that this far exceeds this fellow's understanding, he who wastes his time with the shadows and images useful to him for counter-arguments. But as for failing to note that what acts by transition, so that it passes from one thought to another, from prior to posterior, and not concentrating both together at the same time, – [in short], that this acting according to time possesses the prior and posterior as temporal, – this seems to me to be excessively senseless, as is what he writes subsequently. For, not satisfied with insulting the intellect by considering its activities transitional, he says,[155] 'if one were to concede that not even the intellective substances are free from the relation to time, he would at least consider the creator of time to be higher than any time. If, then', he says, 'God creates the heaven directly' (and everything he added to this, lest I write the same things twice), 'since God also knows the periodic return of the stars, that one of them returns to its position first, and another one later, and he wills it to be so, but he also knows that what is future is future, and that the present is present, and that the past is past, then one thing is also prior and the other posterior in divine thoughts and acts of will, and', he says, 'the knowledge by which he knew they would come to pass is prior to that by which he knew they are present, and again, posterior to this is the [knowledge] by which he knows that they are no longer among beings. If, then, it seems to Aristotle that time is immediately manifested by the expressions "prior" and "posterior", then we will not place even the creator of time outside the temporal relation'. Let him say, however, whether he observes the prior and posterior in acts of thought and will taking place by transition, in that having thought one thing earlier, [the intellect] passes next to another thought, and again from this one to another; or is it according to some other meaning of 'prior' and 'posterior'? For if he observes it by transition and motion, he is the one who says that after the activities of the intellect, those of the God who creates time are also measured by time. And that it is rather in this sense that it thinks the prior and posterior, is clear from [his statement that] 'but when they are present again, he knows that they are present, and when they perish, he knows that they have passed'. If, then, when they are present, he then knows that they are present, and once they perish, he
1163,1 then knows that they have passed, and he does not know these things either earlier or later, but the past, the present, and the future are not simultaneous, but according to flowing time one no longer exists, another exists, and another does not yet exist, it is clear that this fellow imagines that this flow is present in the thoughts of God as well. And if 'when' and 'then' are not temporal adverbs, let this fellow say what kind of thing they are to be considered; but if it is according to some other meaning of prior and posterior that he says one [kind of] knowledge is prior in God, and another one posterior (for let us make him

finer in words, if you will, since so far we do not have the strength to do so in deeds), if, then, he says it is prior and posterior in the sense that God has knowledge of what is prior and posterior among things in time, and knows that one thing came into being earlier, and the other later, then this is not terribly absurd. For time is not entailed by knowledge of the prior and the posterior, any more than it is entailed by its divine cause; nor did Aristotle force us to understand such a prior and posterior temporally. Let no one dare, however, to refer the prior and posterior of the object of knowledge to God's knowledge, and accordingly say that the knowledge by which he knew what would come to pass is prior to that by which he knew that it is present, and that posterior to it is [the knowledge] by which he knows that it is no longer among existing things. For the knowledge of these things is simultaneous in him, and it is not the case that one [knowledge] is prior and the other posterior. Let [Philoponus] recall his own arguments, in which he said that God knows even what is subject to time atemporally; but he apparently did not know what he was saying at the time. But if one viewed the prior and posterior as by position or, as one might say more properly in that case, by causal rank or superiority in nature, in God's acts of thought or will, these meanings have nothing to do with Aristotle's argument, but to attempt to give counter-arguments on the basis of the indeterminacy of homonyms is child's play.

In continuation, he requests whoever hears that time did not exist before it was generated, not to understand 'before' in a temporal sense. 'For many temporal expressions', he says,[156] 'do not necessarily signify time, but merely understanding from the expression that time does not always exist'. But the man who says this clearly does not wish to understand what he is saying. For if he requests that this [sense of] prior and posterior not be understood according to time, he should have investigated in what other sense it can be taken, and having found that it cannot [be understood] in any other way, he should have desisted from his inopportune counter-argument. For whoever says that the world did not exist, say, six thousand years ago, what does he say that is different from the man who says that the art of writing did not exist that many years ago? But what is the man who says this saying, other than that it did not exist in the [period of] time prior to those years? Such a [meaning of] 'prior' is not by position, as if he were saying 'there is no pillar prior to this one' (for after all, position belongs to what is situated), nor by nature, as I said before. For this [meaning] also demands that something be prior, when we say that the genus is prior to the species. In general, however, what is 'six thousand years before' has six thousand years later than it; for what is prior is prior to what is posterior. If, then, the later one is time, the earlier one would also be time, or in time. 'But', he says, 'merely understand this: that time does 1164,1
not always exist'. Yet what does that mean, other than that it sometimes exists, and sometimes does not? Will the Grammarian say that

not even the very expression 'sometimes' (*pote*) indicates time?[157] In that case, he is bound to say what on earth else it could mean; for the expression is obviously not meaningless, but sets in motion some knowledge on the part of the listener, and places him in some disposition. This, then, is what the Grammarian said and heard against the first of the aforementioned proofs.

Against the second of Aristotle's proofs, which confirms that time is everlasting from the fact that all the natural philosophers, except for Plato alone, are obviously of one mind about the everlastingness of time and the fact that it is ungenerated, listen to the counter-arguments this fellow, once again, chose to set forth.[158] First, he says, 'it is not because those except for Plato – perhaps five or ten men – agree in saying that time is ungenerated, that they are worthy of being preferred to Plato. For it is not by the numbers of those who have expressed their view that the truth must be judged: if it were, Aristotle, who was the only one to introduce a fifth substance of the body, would carry off the worst vote of all, or at least worse than most, who say that the principles of the world are the four elements, either all of them or some of them. In general, however, if one must follow the majority of natural philosophers, how comes it, as [Aristotle] himself says in the first [book] of *On the Heavens*,[159] that they all say the world is generated, whereas he demonstrates it to be ungenerated?

Third, since one must be more persuaded by those who say consistent things than by those who disagree even with themselves, when Plato says: 'but time was generated with the heavens', he is speaking more consistently than the others, who said that the world is generated, but time is ungenerated, although neither of these has subsistence without the other. One should therefore be persuaded by Plato, when he says that along with the world, time is also generated. In addition', he says, 'if he proved that in every other area of natural philosophy, the other natural philosophers were led far astray, how can we accept as reliable their testimony to the effect that time is ungenerated?'. And in addition to this, he says he is able to point to many more ancient [philosophers], reliable in many respects, who oppose those who say time is ungenerated.

Yet it must be said against these counter-statements of his as well, that Aristotle does not adduce the testimony of others as demonstrative, but usually after the demonstrations, if he finds some of the previous [philosophers] saying things consistent with what has been demonstrated, he makes mention of them, first of all as an act of gratitude for being the first to see such a truth, and then as offering persuasion to learners, along with demonstrative necessity. For what is agreed on by the majority and the wisest seems more indisputable, as does what is completely undoubted by anyone, since it is not possible to divert the understanding in some other direction, toward the opposite.
1165,1 But this Grammarian, failing to perceive such a goal of these arguments, argues as if against a demonstrative testimony, saying[160]

'one must not follow their undemonstrated assertions, like people bereft of reason'. But that Plato did not call the world and time 'generated' according to the same meaning as Aristotle, and how Aristotle, maintaining an ancient usage, is arguing not against Plato, but against those who understand 'generated' according to the surface interpretation, I do not need to repeat, having clarified these matters, I believe, shortly before in the exegesis of Aristotle's arguments. Thus, even if Aristotle corrects some of the other natural philosophers for saying the world is generated, it is their surface meaning he is objecting to as well. And what is surprising [about the fact] that Aristotle considered that some understood the 'generated' of Plato and the others in the obvious sense, when after so many exegetes of Plato who have articulated how 'generated' is said in Plato, this Grammarian considers that Plato is consistent with himself when he says that the world and time are generated in the sense he [the Grammarian] wishes, saying that they did not exist previously, but came into being later? Nor did he listen to Plato when he said that God, wishing to render the world similar to the model in respect of everlastingness as well, provided the world, instead of the eternity 'there', which provides the model with eternal everlastingness, with time, an image of eternity, granting it temporal everlastingness?[161] I quoted Plato's very words shortly before. But not even when Aristotle introduces the fifth substance does he differ in concept from the others, and especially from Plato. For if the reason why Plato[162] would say that the heavens also consist of fire, earth, and what is in between, is that it is visible and tangible, and nothing is visible without fire, or tangible without earth, we must ask Aristotle, too, if he too would not concede both points: both that the heavens are visible and tangible, and that these [features] belong primarily to fire and to earth. Why, then, does he call it a fifth substance? The answer is that it is because Plato says[163] that the substance of the heavens is different from the four elements beneath the moon, since he attributed the dodecahedral figure to the heavens, while he adorned each of the four [other elements] with a different figure.[164] Thus, he too says there is a fifth substance, that of the heavens; for it is other than the four sublunar elements, since the dodecahedron is the fifth figure, and the figures are substantial. This has been made even clearer by Xenocrates, the truest of Plato's auditors, who wrote the following in his *On Plato's Life*:[165] 'So he divided the living beings again into forms and parts, dividing them in every way, until he arrived at the elements of all living beings, which he called five figures and bodies, into ether, fire, water, earth, and air'. Thus, for him too, the ether is another, fifth simple body, in addition to the four elements.

Nor do they speak inconsistently who say that the world is generated in another sense, either as from a cause[166] or by hypothesis,[167] but that 1166,1
time is ungenerated, not in the sense that not existing previously, they later existed, although time and the world always co-exist. For the

world, owing to the nature of the prior and posterior things within it, received hypothetical generation, while time, possessing within itself both previous and posterior time, could not, even hypothetically, be said to be generated in that sense. For if what is generated is what is previously non-existent, but later exists, but what is previous is a [period of] time, how could one conceive of time as generated? But neither, as this fellow says,[168] is it in all natural problems that Aristotle showed that 'the other natural philosophers have strayed far from the truth', since both here and in many other passages he rejoices in his agreement with them. Did he not expend a great many arguments to prove that almost all the natural philosophers say that generation derives from contraries and some substrate, even though each was led to this concept by a different approach?[169] Nor do I say that all of them declared their agreement in all things, especially with regard to natural realities, but that most often, they agreed in their concepts and differed in the surface aspect of their terminology, against which Aristotle set forth most of his counter-arguments. But since [Philoponus] says he can point to many [philosophers], both more ancient and more reliable, who say that time is generated, did he begrudge us learning who they are, they whom Aristotle was unaware of, since he clearly said they all say that time is ungenerated 'as of one mind except for one',[170] or, being unable to name one, did he really conceal his ignorance by falling silent? But if he points to the legislator of the Jews,[171] saying 'In the beginning God created the heaven and the earth, and the earth was invisible and unformed, and darkness was over the abyss, and the spirit of God was carried over the water', then, once he had created light and divided the light down the middle, and darkness down the middle, he added: 'and God called the light "day" and the darkness "night", and the evening and the morning became one day': if, then, he considers this [generation] of time to be generation by time, let him understand that this is a mythical tradition, drawn from Egyptian myths.[172] In addition, what sense was there in saying that the sun – the cause of the day, as Moses himself says – came into being on the fourth day?

The third of Aristotle's proofs, which says[173] 'if it is impossible for time to exist and to be thought without the now, since nothing can be grasped in time apart from the now, but the now is a mean term, possessing the beginning of future [period of] time and the end of the past one, time must always exist <...> but if there is time, there must obviously be motion as well, since time is some affect of motion'. Beginning to argue against this, he [Philoponus] says:[174] 'I am over-
1167,1 whelmed by amazement that the Philosopher did not see that he has not demonstrated what was sought, but has begged the question. For since the question is whether time is one of the things that always exist, to prove this he made use of the axiom that the now is the mean term of time. But this is the same as to assume that time is everlasting. It is as if someone seeking whether some line, extended to its longest

extension, whose limits cannot be seen, is finite or infinite, were to assume that every point in the line in question is a mean point of the line. For if this is the case, then there is a line before any point one takes, and after it; but if this is so, the line in question is infinite. Thus', he says, 'just as in this example, he who assumes there is a line on either side of every point in the line in question, assumes nothing other than that the line is infinite, and has included the question itself as a premiss of the syllogism, so the person wishing to prove that time is everlasting, then assuming the axiom and premiss that there is time on either side of every now, assumes the very thing that is in question, viz. that time is beginningless and endless'. This is his first refutative argument.

This fellow, as he says, is amazed that Aristotle failed to see that he does not demonstrate, but begs the question. But who could fail to be amazed, and rightly so, that this fellow does not know even the elementary logical methods, but dares to argue against Aristotle about the things that Aristotle himself discovered and perfected? For to beg the question is to prove by itself what is not clear by itself.[175] This occurs when they judge the subject manner to be straightforward and obvious. For instance, when we say that some goods do not immediately make people good, such as capacities, of which good and bad use can be made (for capacity in both directions is good, but it makes good only the person who makes good use of it), the Stoics, confuting this, say[176] 'everything good makes people good', begging the question. Sometimes, some people do not beg the question straightforwardly, but contriving to switch the premisses, they take as self-evident what cannot be demonstrated otherwise than by what is in question. For instance, if they proposed to demonstrate that a triangle has its three angles equal to two right angles, they would assume that the square [has its angles equal to] four [right angles], which is proven by means of the triangle. The third mode of begging the question is when what is being sought is one of two reciprocals, and one assumes the other. For instance, if we sought whether the diagonal is commensurate with the side, one assumed that the side is incommensurable with the diagonal. But perhaps he who assumes the converse proposition is not simply begging the question, since it is from definitions that demonstrations in the proper sense take place, but only when the converse is similarly unclear. Since, then, Aristotle has assumed as self-evident that the now is the mean term of time, and having proved from this that time is everlasting, this fellow says that both are identical. But it is clear from the outset that they are not identical, either in word or in meaning. For the intermediate now is not everlasting time, since the whole of time is made up of both the past and the present. Nor, however, is the now proved to be intermediate by the fact that time is everlasting, but from the concept of the now. But neither is the [proposition that] time is everlasting as clear as the [proposition that] the now is a mean. But if this fellow considers necessary consequences to be identical, it is time to tell him that demonstrations by means of necessary consequences take place through the same things and beg the question, but such are the ones [that take place] by means of essential attributes,[177] and such are demonstrations in the proper sense of the term.

Let us also examine the proof based on a point on a line. Here, too, he took the common feature of the point and the now to be their partlessness, by which they also become beginnings and limits, the former of a line and the latter of time. Yet he failed to take into account that a point can be both the beginning and the limit of a finite line as well, not manifesting the mean, but merely partlessness in things that have position, while the now is not simply partless, but in the manner of the present time. But this immediately implies the past and the future, since it exists neither before nor after, but now. Thus, he who says that there is time on either side of every now, thereby demonstrating that time is everlasting, neither begs the question nor assumes what is under investigation, but assumes what follows necessarily from the self-evident intermediate nature of the now. But who could have expected that someone claiming to be a grammarian would argue against the [proposition] that the present time is always intermediate between the past and the future, and would say, as this fellow does, that the present we experience is like this, but that there once was, and will be again, a present that will not be like this. For he seems to me to talk like someone saying: 'ten is now the limit of the unit numbers, and the beginning of the tens; but there was a time when ten no longer had[178] the unit numbers before it, and there will be a time when it does not have the tens after it, although, *qua* ten, it always has both'. But it is in this way that the present, *qua* present, is the limit of the past, and the beginning of the future. Thus, if this fellow were giving a technical grammatical account in accordance with that now at the beginning which he asserts, he would not include the past tense in verbs. But he would also deny that the imperfects and the perfects, and in general the completed present tenses[179] have any nature at that time, since they have something in common with the past. But this fellow[180] seems to want to conclude to his own hypothesis,[181] and destroys the common notions of names in himself. 'And it cannot be proved', he says, 'that the now is the mean term of time, unless it were demonstrated beforehand that time is everlasting'. Yet if demonstrations must be from what is more evident (but the intermediary nature of the now is far more evident than the [proposition] that time is everlasting: for who does not
1169,1 know, even if he is not a grammarian, that the present time is intermediary between the past and the future?), Aristotle did well to demonstrate everlastingless from the intermediary nature of the now. 'Since, however', he says, 'time came into existence together with the heavens and the world, but none of the arguments proving that the world is everlasting has remained unrefuted, the assumption of the

everlastingness of time will also have been refuted'. If, then, as I consider, I have shown that all his arguments attempting to refute everlastingness are invalid, when examining the first [book] of Aristotle's treatise *On the Heavens*, it is clear that this innovation of his is also empty.

> **251b28-252a3** The same account applies to the fact that motion is imperishable[. For just as in the case of motion being generated, it turned out that there was some change previous to the first one, so, here, [it turns out that there is] one posterior to the last one. For a thing does not cease to be moved and to be movable at the same time – for instance, [it does not cease simultaneously] to be burning and to be combustible, for it is possible for it be combustible but not burning – nor [does it cease simultaneously] to be capable of causing motion and actually causing motion. What causes destruction, moreover, will have to be destroyed once it has caused destruction, and again, later on, what is destructive of it;] after all, destruction is a kind of change.

Having proposed to prove that no motion ever came about unless it also existed previously, nor does it perish in such a way that it does not exist later, he made use of two proofs, one from the definition of motion, and the other from time. He established the question from the latter in two ways: from the prior and posterior, and from the now, which is always in the middle between the past and the future. In the proofs from time, however, he proved both from both: both that there was no [time] when motion did not exist, and that there will be no [time] when there will be no motion. Having proven one of these in the [proof] from the definition, viz. that motion is ungenerated and that there was no [time] when there was no motion, and moving on to the proof from time, he now adds the remaining part of the proof from the definition, viz. the fact that motion is also imperishable. For as in the former passage, he proved that motion is ungenerated by means of the [proposition] that what is capable of motion always pre-exists motion, so in these he proves it to be imperishable by means of the [affirmation] that what is capable of motion always post-exists motion. This is what is meant by 'for [a thing] will not simultaneously cease to be moved and to be movable',[182] nor to cause motion and to be motive. For if what is capable of motion perished as soon as it ceased to cause motion or to be moved, it would not endure after its motion. But if it is capable of being not in motion, but at rest, since nature is the principle of rest, as it is of motion, then a thing will not have ceased to be movable just because it ceases to be moved, if motion is motion in the proper sense, and not the destruction of the thing in motion. For he propounds this argument in the case of motion, as Alexander would have it. If, then, it remains movable once it has ceased to be moved, it will clearly be in motion again, and that motion 1170,1 from which it ceased was not its last motion. And he himself [sc. Aristotle] was satisfied with this, as if it were evident, whereas the Exegete[183] added the reasons as well: for what is motive and movable are either perishable or everlasting. And if they are perishable, even if they perish, every perishing [takes place] through motion,[184] so that once again there will be a motion posterior to the last one mentioned. If they are imperishable, if one is motive and the other movable for infinite time, again, one of them will cause motion, and the other will be moved; for this is what it is to be capable (*dunaton*): that which would occur if what is said to be capable (*dunasthai*) did not perish first.

To confirm that the activities and capacities of things in motion and things that cause motion do not cease at the same time, he adduced the fact that it is not necessary that something cease simultaneously 'from burning and from being combustible'.[185] And by taking 'burning' as his example, he seems to have introduced a lack of clarity into his argument, for [what is on fire] seems to perish at the same time as it is moved, although the argument is true not of perishing things, but of moving things, as the Exegete says. For objects in motion are preserved as they move, and their motion is not perishing. And perishing differs from motion in that what is in motion endures, whereas what is perishing does not endure. How, then, could he adduce 'what is burning', which perishes entirely? The answer is that it is possible to say that he included it instead of 'in motion'; or that it is possible for some things, although they have been burned, not to have perished, like smelted metals such as gold, lead, bronze and so on. Or else it was not in vain that he adduced what is combustible, but to indicate that things moved in the manner of things that are burning and apparently perishing also leave behind something movable after their motion, and something combustible after their burning, which will again be burned. For even if the change of what is burning takes place into fire, since fire also changes into other[186] elements, it too possesses what is combustible, albeit not proximately. But that it was rather according to this concept that Aristotle adduced the combustible, because the argument also holds true of what is perishable, he indicates by adding:[187] 'and what causes perishing will also have to perish', taking the combustible as what is perishable, and proving that the same result will appear from it as well, viz. that there is no last motion, as there is no first one. For since everything in motion is moved by something, so what is burning is burned by something, and what is perishing is destroyed by something. What has caused perishing, then, has not itself perished along with what has been destroyed by it, but, having endured, it will clearly perish later, since it too is perishable; and again, what destroys it [will perish] later. If, then, perishing takes place through motion, and is a motion or a change, as he indicated by saying 'after all, perishing is a kind of change',[188] it is clear that there will be no last change, but it is always possible to identify one later than the [last change] identified.

Thus, motion too[189] is inexhaustible. But why did he assume that what is destructive of something is also itself perishable? For a weak eye is destroyed by staring at the sun, and many plants are dried up by the sun, which has an imperishable substance. The answer is, in the first place, that the aforementioned things are not destroyed by the sun, but 1171,1 by the disposition in the air that is generated from the sun. For if what perishes does so from contraries,[190] and contraries fall under the same genus,[191] and are in the same substrate, and are in every respect of the same kind, then things in generation could not be destroyed proximately by the sun. Next, if the things that are generated are not generated proximately by the sun, but through intermediaries (for the sun generates man, but by means of man[192]), then they would *a fortiori* not be destroyed [by the sun] proximately. The change of sublunar things into one another, and things that destroy and are destroyed in this world, which show that motion is inexhaustible, suffice for the argument. I do not approve, however, of what Alexander says in this passage, in which he says 'in that destruction takes place by something contrary or by something similar, as in the [destruction] that takes place by exhaustion and extinction'.[193] He seems to be saying this of the case of a minor flame that is exhausted[194] and extinguished by a greater one. But it is exhausted and extinguished not *qua* a flame by a flame, but *qua* something lesser by something greater, which again are contraries. It is worth noting that Aristotle, having proved that motion is ungenerated, and having been able to prove from the fact that it is ungenerated that it is also imperishable, since no one so far has been bold enough to say that something ungenerated is perishable, did not think it well to carry out his demonstration on this basis, since he has not yet even proved that these things reciprocate – viz. the generated and the perishable, and the ungenerated and the imperishable[195] – which he demonstrates later, in the first book of the treatise *On the Heavens*.

252a3-5 But if this is impossible[, it clear that there is an eternal motion, and not that it sometimes existed] and sometimes did not.

Having demonstrated that it is impossible for motion to be generated or perishable, the former from the fact that every motion taken as the first is necessarily pre-existed by another motion, and the latter from the fact that another [motion] post-exists each one taken as the last, and having also proved these things from the everlastingness of time, which necessarily accompanies motion, he rightly said in conclusion that if these things are impossible, viz. that motion should be generated and perishable – that is, that it should possess a beginning or an end, which is the same as there being a first and last [motion] – then it must be everlasting, and not be such that it sometimes exists, and sometimes does not exist.

The Grammarian,[196] arrogant as usual, says: 'if the [arguments] by

means of which the Philosopher undertook to demonstrate that motion is ungenerated are the same as those by which he establishes that it is imperishable, then', he says, 'it is by the same [arguments] we used to dismiss those arguments, that we correct the present ones'. It must be said against him, then, in accordance with what the more recent commentators call the 'inference by three',[197] that if the arguments by means of which this fellow attempted to overturn the [proposition] that
1172,1 motion is ungenerated, having been sounded out on all sides,[198] have appeared to be invalid, although stated in all kinds of ways, [then] those that correct the imperishable character [of motion] will undergo the same thing when put to the test. But since, as if he had won the first fall, [Philoponus] next does not hesitate to brazenly make abundant use of less substantial arguments, let us set them forth, too, and examine them. 'In the case', he says,[199] 'of things whose being is in motion, or simply all those things whose natural motion maintains and perfects their being, these things, once they have ceased to be moved, also cease to exist. Indeed, the heart, the bronchial tubes, and the lungs, as soon as their motion is destroyed, have their capacity of moving destroyed along with them, or rather the very fact of being what they are called, and [this is true of] each of the other parts of the animal. And fire', he says, '[exists] as long as it is in motion, but once it stops moving, it also stops being fire. It is therefore not necessary', he says, 'after all the postulated destruction of the last motion, that another motion remain behind, posterior to the last one'. Here, too, however, will this fellow seem to understand what is said, when Aristotle says 'for it does not simultaneously cease to be moved and to be movable',[200] that is, capable of moving, not necessarily by the same motion, but by whatever kind [of motion]? For his purpose was not to prove that the motion of each thing in generation is everlasting, since he does not even say that their substance is everlasting, but that motion always exists in sublunar things as well, and [that] there is no last motion, but there is one after this one, too, although one [motion] always succeeds the other. Yet this fellow says the heart, fire, and other such things, cease to exist once they have ceased to move, failing to understand that they cease to be heart and fire, and to have the capacity by which they will be moved by the same motion, but not to be movable. For when the heart perishes, it is dissolved – into the elements, that is, from which it was composed – and fire changes into air. But I do not think even this fellow doubts that the elements, especially air, are movable; and not only that, but they are in motion. Observe how, although Aristotle has refuted such madness beforehand, by saying 'for instance, burning and being combustible'[201] this fellow did not have the ability to understand what was said, but he considered [Aristotle] said that once motion has ceased, the very same thing remains, movable by the same motion, although he should have known better, if only from the comparison with generation. 'For', he (Aristotle) says,[202] 'just as in the case of motion coming into

being, it resulted that there was some change prior to the first one, here, in the same way, there is [a motion] posterior to the last one'. Thus, as in the former case it was proved that prior to the [motion] supposed to be first was [the motion] by which what was first moved was either generated (if it was generable), or (if it was everlasting), changed into [the state of] being capable of motion, so here, posterior to the [motion] supposed to be last, there is another [motion], by which what is left over will be moved, or by which what has caused destruction will be destroyed. But the Grammarian has obviously not been able to follow either the former or the latter considerations.

But let us look at his next argument, which seems to be even more clever: 'Yet even if it is true', he says,[203] 'that after motion has ceased 1173,1
something remains that has the capacity for being moved, not even in this case does the Philosopher correctly conclude what follows. For if not everything that comes into being does so through motion, but there is, according to Aristotle, a generation [that takes place] all at once (*athroa*), without motion and temporal extension, then there is also a destruction of the same kind, such as the presence of perfect forms in their substrates, and their withdrawal from them, and like points come to be united, and like contacts, and lightning, and the apprehension of optical sensation.[204] Therefore, not everything that perishes does so through motion'. Thus, he would be saying – for he seems to me to have left his argument without a conclusion – that even if what causes destruction perishes, it does not necessarily perish through motion. 'And if God the Demiurge', he says, 'produces without temporal extension the heavens and the world, produced directly by him, then when he should wish to destroy the world, its destruction', he says, 'will also be non-temporal'. This fellow said these things in the hope of establishing that not everything that perishes does so through motion. Here too, however, he should have listened to what is said by Aristotle, when he says, 'after all, destruction is also a kind of change',[205] and in book five of this treatise, where he says: 'Every change is from something to something. This is shown by the name as well, for it indicates that one thing is after another, and one is prior, the other posterior',[206] and that generation and destruction are changes in substance. Having heard these [statements], [Philoponus] should have understood that when Aristotle speaks of some generation, or destruction, or alteration that is 'all at once',[207] he does not say this in the sense that it is generated without change and time (for something comes after something else in these as well, and there is prior and posterior and extension in them, and transition from not-being to being, and from being to not-being), but in these things extension, change, and time are concentrated.[208] Thus, their instantaneousness seems to accrue to them not when it is observed in them throughout the entire change, but at its end (*telos*). For who would say that the construction of a house is timeless? But the end (*telos*) of the form, although not the form in the simple sense, is

provided by the placing of the last roof-tile, just as the beginning of its destruction is the removal [of that tile]. Thus, the form is said to be present and to depart timelessly. The division and unification of a line, too, and the other [examples] adduced, could not have occurred unless some things had previously been moved by some motion, at the limit of which these things come into existence. And if someone, looking at the manifest limits, considered them to be timeless, he would also say that animals and plants come into being timelessly, characterising motion by its end. Yet it is one thing to come into being (*to ginesthai*), and another to have come into being (*to gegonenai*); and generation is observed primarily in coming into being, not in having come into being, and motion is [observed] in being moved (*en tôi kineisthai*), not in having moved (*en tôi kekinêsthai*).[209] But if, in some things, the extension of generation and alteration is more concentrated, as is said in the
1174,1 case of lightning, contact, and the curdling of milk, and [the freezing] of water, in which the change is said to be all at once, 'all at once' is not said in the sense that it is timeless, but in the sense that it takes place as a whole (*kath' holon*), and not part by part (*kata meros*). In general, one must not consider that something can have come into being (*gegonenai*) without [the process of] coming into being. However, 'having come into being' (*to gegonenai*), too, if it is not merely a meaningless term, indicates having come into existence through generation, and 'having perished' (*to ephtharthai*) means having travelled through perishing.

But perhaps this fellow concedes the [argument from] generation and destruction, but does not agree with the one from motion, failing to understand that the motion Aristotle says is everlasting in this passage is that which is predicated in common of generation and destruction, and amounts to the same thing as change. This is why he said, in this sense, 'after all, destruction is also a kind of change',[210] and before that,[211] 'for just as in the case of the coming into being of motion, it resulted that there was some change prior to the first [motion], so here there is [a motion] posterior to the last [motion]', using the words 'motion' and 'change' indifferently. Rather, he was indicating that even if what is supposed to be first or later is one species of motion, and what is before the first one and after the last one is some other species of change, motion is everlasting in this sense as well, even if it comes into existence in various species. But if this fellow agrees that the Demiurge produces the heavens and world timelessly, how can he say that he produces [them] later after not having produced [them] previously, and after having produced [them], he later ceases to produce [them]? For he does not consider that [God] is producing now, already having produced, and still after a certain number of years he destroys them. How, then, does he produce and destroy the world timelessly, although he acts on it in accordance with the extension of time, since he sometimes produces and sometimes destroys, and he produces some things earlier,

and others later, as this fellow agrees, and sometimes these things, sometimes those other things, so that having created the world in six days, he brought about something different every day?[212] For as long as the meaning these [terms] exhibit is temporal, and no other, it is not possible for he who acts according to them not to act temporally. But since he produces atemporally – for this is true – he always produces, and not sometimes yes, sometimes no; for these [expressions] are temporal. But this fellow does not notice that he is combining the incompatible, viz. what is timeless and what sometimes exists, but sometimes does not. And yet, this point alone, which this fellow concedes, sufficed to prove the everlastingness of the world: the fact that God produces it directly, and produces it timelessly, and not through any intermediary. For if he produces it timelessly, it is not the case that he sometimes produces, and sometimes does not. But if so, then it is clear that he always produces. But [Philoponus], as if it were already granted to him that God sometimes produces <the world>[213] and sometimes does not (in other words, in a part of time) timelessly, says 'and when he should wish to destroy the world, he destroys it timelessly'. If, then, 'when' also signifies time, he would again be saying that [God] destroys it timelessly in a part of time, and that,[214] if he creates in a part of time, he must create some things first, others later, and some things in one [period of] time, and others in another. For how is it possible to 1175,1
make the wandering stars inhere in all the places of the zodiac at the same time? And whence did he make each one begin its motion, or what position did the heavens have? For if he first created spring in our climes, then ceased his creation, having created in six days or however many it was, it is clear that he will not be the creator of summer and winter, according to them. The cause of this puzzle, and others like it, is considering that God acts similarly to men, and granting timelessness in name alone, postulating that his activities are frankly temporal, saying that he sometimes creates, sometimes does not create, and sometimes destroys, sometimes does not destroy, whereas they should see, when they say that he creates timelessly, that he must always create.

Next, then,[215] [he says] – confusedly, as it seems to me –, contradicting Aristotle when he says, 'and what is destructive will have to perish once it has destroyed',[216] and against Alexander, who says 'but he assumed that what is destructive of something is itself perishable, perhaps because destruction comes about either from a contrary or from something similar', and that what destroys, if it remains everlasting, will again be moved, but if it perished through motion, it will perish.[217] Arguing against these considerations, then, [Philoponus] says he has shown in the fourth discourse[218] that what comes into being does not necessarily do so from its contrary, nor does what perishes change into what is contrary to it, nor does it undergo destruction from its contrary, since it is not even the case that there is something contrary

to everything that comes into being and perishes. Instead, some of the things that perish do not have an external cause of their perishing, but because their natural capacity is measured by time, they grow ill and waste away bit by bit, even if nothing harms them from outside. '[Aristotle] should not, therefore', he says, 'have asked, in the case of everything that perishes, whether the cause of destruction endures after what has the capacity of destroying has caused destruction, and whether it is everlasting or perishable, and thus he reduces the absurdity to the infinite, by means of the heap argument'.[219] Against this, that what is generated – and according to Aristotle this is, as has often been shown, what later exists after having not existed earlier – that these things, then, are generated out of what exist, either from contraries or from their own privations, and perish either into contraries or into privation, but again into existent things, has been shown primarily in what was said in the [commentary] on the first book of the *Physics*, but it has also been shown in the passages where I tried to put to the test what this fellow boldly said against the first book of *On the Heavens*, sounding out his fourth book, which he mentioned here. But appropriate discussion was devoted to these matters shortly above,[220] although we must touch upon the argument here as well. If things exist later after having not existed previously, it is impossible for them to be generated out of what does not exist; for what is generated out of existent things manifests itself in accordance with the series of what
1176,1 has preceded and of what follows. The Trojan War, for instance, was capable of occurring then and not at some other time, through the series of the things preceding those realities, constituting them, and following them. As for the things that come into existence out of what does not exist, why should they come about in this [period of] time, rather than in that one? Not to mention that comings-into-being and destructions are clearly kinds of change, with one thing receiving its subsistence after another.[221] If, then, what comes into being does so out of something, it comes into being either out of something similar, or out of a contrary, or out of its own privation. But this fellow obviously rejects [generation] out of what is similar; it therefore [comes into being] either out of a contrary or out of privation, since there is nothing contrary to substance.[222] Thus, in the case of things that have no contrary, coming-into-being derives from privation, although in this case too, to the observer, comings-into-being and destructions take place by the change of contraries into one another. For when flesh comes into being out of bread (which is not flesh, but is aptly disposed (*epitêdeiôs ekhei*) to become flesh), the change takes place in accordance with the mutual action and passion of contrary qualities. 'But', he says,[223] 'it is not necessary that the cause of perishing be external, but sometimes the measured aspect of the capacity of perishing things is sufficient for wasting away, even if there is nothing to harm them'. Once again, then, the same argument will appear, and since this fellow was unaware of

it, he was forced to say and hear a great many things. After all, if the wasting away took place toward what does not exist in any way (*eis to mêdamêi mêdamôs on*), the substrate would soon waste away as well, if nothing impeded it. But if it must change from cold to hot, it would clearly not change into the contrary by itself, but what was until then cold, but potentially hot, is changed by what is hot in act. But how can it also be possible for all things in the world of becoming, since they are made up of contrary qualities and approach one another, to fail to act upon one another, or to undergo effects from one another, and yet their capacity wastes away without impediment? For my part, I believe that the measures of capacity contribute to the action and passion of contrary qualities. Accordingly, the same thing perishes more quickly when it encounters certain associations, but less quickly when it has others. Against this fellow's argument, I would say that if this active and passive aspect were not what destroys the qualities in the world of becoming and is destroyed by them, but were only imparting and partaking perfectively and vitally,[224] as in celestial things, the capacity of their substance would not be wasting away, nor measured by a part of time. The Exegete was therefore also right to suppose that what destroys is external to what is destroyed, and whether [the destructive agent] is everlasting or perishable, he concluded that there will be motion after the [motion] supposed to be last. But if we must take notice of this, too, for the sake of those who are deceived, without examining them, by the multitude of this fellow's arguments, then it seems to me that this fellow considers that he can reduce the heap argument to an infinite regress, and I think this is sufficient to prove that he is not even aware of the difference between them. For the reduction to infinity, which is demonstrative, never has any cessation in its progress, as if one were to say that everything that causes motion is itself moved, and everything that is moved by something is moved by something external. 1177,1
For once these [propositions] are supposed, one must continue to infinity, postulating one motive cause before another. The heap argument, in contrast, is a sophistical argument which concludes, on the basis of what they call the 'little by little' interrogation,[225] by means of the weakening of representations, to what is unclear or manifestly false. For they ask whether the first drop had an effect on the stone, or not. For if it did, why can't this effect be seen? If it did not, then neither did the second or the last one. How, then, did the hollow in the stone come about? You can see that this [argument] does not reduce to the infinite; how, then, can he say that there is reduction to the infinite by means of the heap argument? But let this be written for those who wish to judge this man's condition. In general, what need did he have of the heap argument here, unless it were to put on a terminological display? But Alexander said that the things that perish do so with some things being in motion, viz. the things that cause them to perish (for this is also what Aristotle said), and if these things are indestructible, they will be in

motion again, and if they are perishable, then even in this case there will once again be some [motion] posterior to the last motion, [viz.] the one that occurs on the occasion of their perishing. This fellow, failing to understand, considers that what Alexander here calls 'things in motion' are not the things that cause perishing, but the things that are perishing, he says[226] 'as, for instance, the elements, which are themselves perishable, again themselves perish when other things are in motion, such as matter and form; but matter, no longer perishing along with realities, remains in possession of the capacity to be in motion'. Thus, this fellow, misunderstanding and not even heeding Aristotle, when he says,[227] 'and what is destructive will have to be destroyed once it causes destruction, and again, what is destructive of it [will perish] later', says he has solved this puzzle on many grounds. 'For', he says, 'even if nature does not create prime matter, but God creates it not out of matter, as He destroys it, when he wishes, into the not-being from which it came into being, just as', he says, 'form does not change into another form, but reverts into utter not-being, whence it had come into being'. But it has been proved many times, I believe, that the things produced by the Demiurge out of what does not exist are not generated in the sense of 'generated' according to which what does not exist previously exists later, but that their subsistence is everlasting. For there is no reason why the things produced out of what does not exist should arrive among existent things at that moment and at no other time, as the things that are generated out of existent things have an easily detected cause for coming about at that moment: because, as I said, of the series of their antecedents and consequents. But these are necessarily everlasting, whereas generated things, in the aforementioned manner, have their generation out of what exists, and their corruption into what exists, since generation and corruption are change. And what form changes, not from what was previously potential to what is in act, and from what is in act to its own privation, but returning to utter not-being, as this fellow says?

Having said this,[228] he promises to prove that the change of the world takes place not into complete non-existence, but into something greater
1178,1 and more divine. And it is amazing that he thinks the destruction of the universe is a change into something existent and more divine, but says that it does not have its generation out of what exists. But he indicates, as he elaborates in the following books,[229] that this world changes into some other more divine world, not noticing that this is not the destruction of the world, but its perfection. Concluding his proofs against the arguments showing that motion is ungenerated and imperishable, he says they have been sufficiently refuted by his prattle, which is so invalid. And as if he had refuted those [arguments], he dares, on his own, to demonstrate that it is impossible for motion to be ungenerated. He assumes beforehand three axioms for his demonstration: first, that if one of the things that come into being necessarily requires something

pre-existent in order to be generated, as a ship requires wood, it could
not be generated unless those things had come into being beforehand.[230]
Second, that it is impossible for an infinite number to exist in act, or for
someone to traverse the infinite by counting, and that it is impossible
for [one infinite] to be greater than the infinite, or to increase it. Third,
that if for the generation of something it was necessary for an infinite
[number of] things to pre-exist, one being generated out of another, it 15
is impossible for that thing to come about. 'For on this basis', he says,
'Aristotle too proved in the *On Generation*[231] that it is impossible for the
elements of bodies to be infinite in number as well, since they are
generated out of each other. For the infinite cannot be traversed, so that
fire, for instance, will not manage to come into being, if it could do so
[only] after an infinite number of things had come into being previously.
These things having been agreed beforehand, therefore', he says, 'if the 20
particular motion of this fire has a beginning and limit to its being, but
in order for it to come into being, another motion had to occur before it,
by which, from changing air, for instance, the body of fire had to come
into being; and again, before the motion of air, which has changed into
fire, another motion pre-existed, that of water, for instance, by which it 25
changed into air, and another before this one, and so on to infinity: if
neither the world nor the change of realities into one another had a
beginning, then an infinite [number of] motions must have pre-existed
in order for this particular fire to come into being. For it would not have
come into being, if an infinite [number of them] had not pre-existed',
because of what he says is the first axiom. 'If, then, it is impossible for
an infinite [number of] motions to have occurred in act, because of the 30
second axiom, then it was not even possible for the motion of the
particular fire to come into existence, because of these and because of
the third axiom, which says that what requires an infinite [number of]
things pre-existing its generation could not come into being.[232] If, then,
the motion of the particular fire has come into being, then an infinite
[number of] motions could not have pre-existed it, according to the
necessity of conversion with the antithesis'[233] (so this fellow says, he 35
who does not even know what conversion with the anthithesis is, as I
believe I have shown when putting to the test what he says against the
first book of the *On the Heavens*).[234] 'If, then', he says, 'nature went
through a finite [number of] motions in order to create the motion of
particular fire, then there is some first motion, not pre-existed by any 1179,1
other, and the same reasoning holds true of the other individual mo-
tions, this too', he says, 'is in accord with the natural argument. For
complete things are primary with regard to incomplete things, and
things in act are primary with regard to potential things. If, then, the
ascent were to take place *ad infinitum*, neither will complete things
precede the incomplete, nor will things in act precede what is potential, 5
but if [the motions] are finite, the first one, since it will of course have
come into existence at the same time as the universe, will have made a

beginning of its subsequent motions from what is in act and is complete. But if', he says, 'there is some beginning of particular motions, and it is not possible to conceive one motion before another *ad infinitum*, then the circular motion of the celestial bodies must also have begun from some beginning, having not existed previously. For the heavens cannot always be in an identical state, while the things within it, generated and perishable, have[235] their being in coming to be and in perishing'. He also adds this fourth [point]: 'If the motions that are to come, when added to those that have come to be, increase their number, but the infinite cannot be increased, the motions that have come into being previously cannot be infinite'. Fifth, he says: 'If the spheres do not move with equal speed, but one does so in thirty years, another in twelve,[236] and the next ones in less time, as the sphere of the moon [completes its orbit] in one month, and that of the fixed stars in a day and a night, but the motion of the heavens in beginningless, then the sphere of Saturn must have revolved through infinite revolutions, that of Jupiter almost three times as many, those of the sun will be thirty times those of Saturn, those of the moon three hundred and sixty times, and those of the fixed stars more than ten thousand times as many. How, then, could it not be beyond every absurdity, if it is not even possible for the infinite to be traversed once, to postulate something ten thousand times greater than the infinite, or rather to postulate something infinite times greater than the infinite. Thus', he says, 'it is necessary that the circular motion of the heavenly bodies begin to exist after having not existed previously, when the heavens themselves (as he says, for this is not what I say) began to exist'.

In these considerations, it is possible that the axioms have been assumed truly, but because he was unaware of one, I believe, this fellow heaped up a huge pile of counter-arguments, misleading himself in the first instance, and then those who believe that the destruction of the world confirms their own doctrine. What this one [axiom] says is this: what is infinite in act and in the proper sense of the term, whether it is supposed to be infinite in number or in size, is not infinite in one aspect but finite in another. For such a thing is no more infinite than it is finite, but what proceeds to infinity (*ep' apeiron*) neither is nor is said to be infinite in the proper sense, either. For what proceeds to infinity has not yet assumed infinity, and even if it never ceases proceeding toward it, it will never assume infinity. But this fellow, assuming the 1180,1 axioms as applying to things that are infinite in act, in the case of which it is true that it is impossible for an infinite number to come into existence, or for someone to traverse the infinite by counting, or for the infinite to be increased, or for a number infinite in act to come into existence prior to the generation of a thing, as Aristotle proved in the case of the elements, he no longer applied[237] his demonstrations to this kind of infinite, viz. the kind that exists all at once, but to what comes into being *ad infinitum*, as we say motion is, failing to understand that although we say motion is beginningless and endless, the [motion] that [extends] down to now, at any rate, is not infinite, for we have supposed it to have a limit, viz. the now. Thus, a motion infinite in act does not pre-exist the motion of this particular fire, for the [motion taking place] until now is limited by this now. Nor is this fire here prevented from coming into being through pre-existent motions, for they are finite down to what has occurred. And not only that, but the existing causes of the fire are also finite, since the four elements are generated out of each other, revolving in a circle: for fire comes out of air, air from water, water from earth, and earth from fire; and again, fire comes from air, and so on *ad infinitum*. But he should have learned from Aristotle the nature of the *ad infinitum*, and how it differs from the infinite in act, and then raise some puzzle against it, if he could, instead of applying to the *ad infinitum* the absurdities concluded in the case of the infinite in act. But Aristotle says the following about this kind of infinite, viz. the *ad infinitum*, in the third book of this treatise:[238] 'Such, in general, is what is infinite by being always taken again and again, and what is taken is always finite, although it is always other'. In addition, being is said in many senses, 'so that the infinite must not be taken as a concrete thing, as a man or a house, but as a day or a contest is said, whose being did not come about like a substance, but is always limited in generation and destruction, but always other'. Thus, even if my father exists before me, and his father before him, and we could trace this back *ad infinitum*, as if there were no beginning, the things that came into being previously are not infinite, for they have so far been limited as far as me. Yet they have not stopped, but one succeeds another, always finite.

Such, then, is the generation and destruction of individuals, which never fails, but always reaches a limit, owing to the nature of the *ad infinitum*. But the Grammarian, having assumed his axioms in the case of the infinite in act, [and assuming] they hold true of what does not even subsist as a whole,[239] as he himself admits, says 'if an infinite [number of things] must pre-exist for the generation of a thing, one coming to being out of another, then it cannot come into being, since it was necessary that an infinite [number of] motions pre-exist it'. He clearly applies the impossibility that follows from the infinite that is in act altogether as a whole, and is infinite in all directions, to the *ad infinitum*, in which everything that is taken necessarily exists, but is always other. For if the things that came into being previously were infinite in act, truly nothing would come into being out of them, not only because nothing exists outside the infinite, but because it is impossible for what comes into being to be joined to the infinite, which has neither beginning nor end. But Aristotle in book two of the *On Generation* – since this fellow mentioned this passage as well, having understood nothing of what is said there – says that from the procession of things in generation and destruction, supposed to be *ad infinitum*, it does not follow that the elements have not come into being, as this fellow

thought, since he was not even capable of understanding that this is precisely Aristotle's intention in this passage: to show that the change of the elements into one another, as it comes round in a circle, is *ad infinitum* and perpetual, but not by supposing that the contrarieties specifying the elements are infinite in act. This follows for those who say that they come round formally in a circle not simply *ad infinitum*, as we say, but that the things that come into being exchange their forms *ad infinitum* as they proceed in a straight line, as for instance air from fire, and water from air, and earth from water, and something else from earth, and again something else from this *ad infinitum*, differing formally in this way. If this is so, however, it is necessary that the contrarieties specifying the elements, according to which change takes place, be infinite in act in the nature of beings. For instance, since there would be no change from white to black if black did not exist among beings, so it would not be possible to exchange forms *ad infinitum*, unless forms were infinite in number. For if they were finite, it would be impossible for the exchange of forms to take place *ad infinitum*, for they would run out. Thus, from the formal change of the elements, supposed to be *ad infinitum*, it would follow that the contrarieties would be infinite in act, which does not follow for us. For we say that generation and motion proceed *ad infinitum*, but that that there is no exchange of forms, but they go round in a circle. This fellow wants to introduce this as well, by eliminating the formal progression *ad infinitum* in a straight line. But he eliminates it by saying that it follows from [such a progression] that the formal contrarieties are infinite in act, and that absurdities follow from this, one of which eliminates our knowledge concerning beings, and the other generation. For the knowledge of the essence of each thing (*to ti estin*) is present in us through definitions: but if the specific differences by which the elements are specified are infinite in act, if the exchange of elements is infinite in a straight line, with one form being generated out of another *ad infinitum*, as I said, then definition is impossible,[240] since definitions derive from the genus and from determinate differences. But neither is it possible for anything to come into being if the intermediaries are infinite in act and untraversable; for if these are finite, then the extreme terms must also be finite. Thus, that infinite which is supposed to be a simultaneous whole (*hama holon*) is one thing, and the *ad infinitum*, which has its being in coming-into-being and perishing, is another, as has been said. And what is taken of the latter is always finite, but ever different, and [the process of] generation is not hindered
1182,1 by this, but rather cooperates with it. Thus, the motion of the particular fire, and of the other things, did come into being (*gegonen*); and an infinite [number of] motions – which are indeed now finite – did not pre-exist them, and there is no first motion, but one can always take [a motion] that is earlier than the one supposed to be first. But if this is true, the circular motion of the celestial bodies, from which the inex-

haustible motion of sublunar things is preserved, must also be beginningless. But what the added motions increase is not the number, infinite in act, of the motions, as this fellow says, but the finite [number]. For the [number of motions] until now is finite, and each one that is added is finite, not infinite, even if they are added *ad infinitum*. Thus, even if the spheres move with unequal velocity, and the return of each one to its initial position is beginningless, those that have taken place until now are not infinite, nor will they ever become infinite, prevented as they are from becoming infinite by the very fact that they proceed *ad infinitum*. For what is *ad infinitum* always has something missing (for potentiality always inheres in it), while the infinite, if there were such a thing, would be lacking nothing in act. But neither does what [Philoponus] says agree with the natural account, which says that perfect things precede the imperfect, and that what is in act [precedes] what is potential, when this fellow does not even seem to know in what cases this is true. For it is true of wholes, but in [the process of] generation it is more in the case of the same thing that what is imperfect [pre-exists] what is perfect, and that what is potential pre-exists what is in act, even if, as a matter of fact, they derive their generation out of each other; after all, man comes from sperm, and sperm from man. And I have said this with regard to temporal pre-existence, which seems to be the only kind this fellow is familiar with; since at least in capacity, causality, and superiority of value, what is perfect necessarily pre-exists what is imperfect, and what is in act [pre-exists] what is potential, and what is active [pre-exists] what is passive in time as well. In [the process of] generation, however, as I said, since there is no temporal beginning, one could not say that what is in act simply temporally pre-exists what is potential; for what is potential in each thing temporally pre-exists what is in act in it, as the egg precedes the bird.

Let this be said, then, against the Grammarian's sixth book, which sought to eliminate the everlastingness of motion, lest, by its means, the world might be demonstrated to be everlasting. For he set himself one goal: that those who have the misfortune of being persuaded by him should not consider either that God is unchanging, or that the world is everlasting. Having proceeded as far as the aforementioned remarks, however, he considered that he had completed his counter-argument against the everlastingness of motion, unable as he was even to notice that the key points of the demonstration are transmitted in what follows, where Aristotle proves that of beings, some are immobile, some always in motion, and others sometimes in motion and sometimes at rest. But let us now return, if you will, to Aristotle's continuous text, now that we have freed those who encounter the material taken up previously from that uneducated yapping.

1183,1 **252a4** Indeed, speaking in that way seems rather to resemble a fiction.

Having shown that motion is everlasting by means of the everlastingness of time, and that it is impossible to suppose motion or time to be coming into being or perishing, he now says that motion and time are also proved to be everlasting from the fact that not to speak in that way, but to say that motion or time sometimes exists and sometimes does not, 'rather resembles a fiction'. For what could be more fictional than to say that motion and time, not having existed previously, then came into being? For why does (the fact that they existed) then, and not earlier or later, rather resemble a fiction? Perhaps he means that it is fictional to say that time existed when there was no time; but this is what is said by whoever affirms that sometimes there was time, and sometimes not. But the same fiction applies to motion as well, since time is the number of motion; for if time existed when there was no motion, but when motion existed, time did not even exist, it is clear that whoever says this is saying nothing other than that there was time when there was no time.[241] 'He would be saying', as Alexander says, 'that to speak like this rather resembles a fiction, and therefore the people who said this spoke without demonstration or reasonable arguments, or because since motion has been proven to be immobile,[242] it is unpersuasive to call it destructible'. Unless it were because [Aristotle] was saying 'fictional' not merely with regard to destructibility alone, but it was also given with regard to its generability.

252a5-19 Similarly, to say that things naturally tend to be that way[, and that this must be considered a principle, which is what Empedocles seems to say, that the fact that Love and Strife dominate and cause motion alternately belongs to things necessarily, but that they rest for the intermediate period. Perhaps those who postulate one principle, such as Anaxagoras, would also speak this way. However, of the things that are by nature and according to nature, none are disorderly, for nature is the cause of order for all. The infinite, by contrast, has no proportion to the infinite; yet all order is proportion. For something to be at rest for an eternal time, then be moved at some moment, but for there to be no difference whether this takes place now or before, nor do these things have any order: this is no longer the work of nature. For what is by nature either behaves simply, not sometimes this way and sometimes that, as fire is by nature transported upward, not sometimes thus and sometimes not,] or what is not simple has proportion.

Having said that to say that motion sometimes existed and sometimes did not 'resembles a fiction', since no cause for this can be given, he goes on to say that to say that things 'tend by nature to be this way', giving this as a principle and a cause, is equally fictional. But this is what Empedocles would seem to be saying, when he says that 'Love and Strife reigning and moving in turn belongs necessarily to realities',[243] and if this is true, then it is also true that they are at rest in the intermediary period, for between contrary motions there is rest. Eudemos[244] understands immobility in the reign of Love as the Sphere, when all things are combined,

> there where the fleet limbs of the sun are not discerned,[245]

But, as he says,

> thus, in the dense darkness of Harmonia is fixed
> circular Sphairos, proud in his ever-joyful solitude. 1184,1

But when Strife has once again begun to reign, then motion again comes about in the Sphere:

> For all the god's limbs began, one after the other, to tremble.[246]

But what difference is there between saying 'because things tend by nature to be that way', and [saying] 'by necessity', without adding the cause? This is what Empedocles seems to be saying:

> but they reign in turn as time goes round,[247]

and when he blames necessity for the things that occur:

> Necessity is an oracle of the gods, an ancient seal,
> eternal, and secured with broad oaths.[248]

For he says that it is through necessity and these oaths that each of them reigns in turn. But Empedocles also says the following of the reign of Strife:

> But when great Strife prevailed[249] in its members
> and rose up to honour in the completion of time,
> which was traced for them in exchange for a broad oath.[250]

These things, then, which are said without a cause, [Aristotle] says are no different from saying that [things] 'tend by nature to be that way'.

But perhaps, he says, Anaxagoras too, who says the intellect is the one principle of motion of the homoiomeries that had previously been at rest, and does not add the reason why it set them in motion at that time, but not previously, is saying nothing other than that fictional 'because [things] tend by nature to be that way'. Yet what is without cause or order is to be avoided, especially in the case of nature and the things that come about by nature, since nature, as a proportion (*logos*), is the cause of order and proportion (*logos*) to all things. But the [period of] time during which [intellect] was previously at rest, being infinite with regard to the [period of time] from which it began to move, has no proportion (*logos*); for what is infinite has no proportion (*logos*) to what is infinite. But neither does it [have any proportion] to what is finite. One concludes, then, in the second figure, that infinite rest does not exist by nature with regard to either infinite or finite motion. And to remain at rest for an infinite time, then to be moved at a given moment, and to state no difference of this, by which they are moved now rather than before, nor do they have any order: this is not the work of nature, for what is by nature is not sometimes thus, sometimes otherwise, without proportion or cause, but everything that is by nature is disposed either simply or[251] similarly; for instance, we say that fire is transported upward by nature, because it is not the case that it is sometimes so transported and sometimes not, but it is simply transported upward universally, and always in a similar way. Earth, too, [is also transported] simply downward. But if one of the things that are by nature is not always similarly disposed, but sometimes yes and sometimes no, there is some proportion (*logos*) and cause for this lack of simplicity and difference. For instance, the cause of the fact that it is not always summer or always winter, but that they alternate, is the inclination of the zodiac and the sun's motion along it; and [the cause of] the fact that animals do not always engender is that being able to create things similar to oneself pertains to what is perfect.

Eudemus[252] reproaches Anaxagoras not only because he says that motion, after not having existed, began at a certain moment, but also because he omitted to talk about whether it endures or will cease, although this is not obvious. 'For what', he says, 'is there to prevent one from opining that all things stop at a given moment by virtue of the intellect, as that fellow said they move?'. But Eudemos also reprehends Anaxagoras for the following point: 'How is it possible for some privation to be prior to its opposing possession (*hexis*)? If, then, rest is the privation of motion, it could not exist prior to motion'.

252a19-252b5 It is therefore better as Empedocles says [, and if anyone else said things are this way, that the all is alternately at rest and again in motion, for such a process already possesses some order. This too, however, the speaker must not merely affirm, but also state its cause, and not merely postulate it nor consider it an irrational axiom, but apply to it either induction or

> demonstration. For what is postulated is not itself a cause, nor was [the state of] being Love or Strife, but of one it was to gather together, and of the other to separate. If what is individual is determined in addition, in these cases it must be stated in such a way that there is something that gathers men together, viz. love, and enemies flee one another: for this is what they postulate exists in the Whole as well. For matters are clearly this way in some cases, but that they are so for equal periods of time requires some rationale. In general, however, to consider that it is a sufficient principle, if something always exists or comes into being in this way, it is not correct to suppose, although Democritus refers the causes concerning nature to this principle, inasmuch as things also took place in this way previously. Of [the fact that something occurs] always, by contrast, he does not think it well to search for a principle, which it is correct to say in the case of some things, but incorrect to say that it holds true in every case. After all, a triangle has its angles equal to two right angles, yet nevertheless there is another cause of this eternity.] Of principles, by contrast, there is no other cause, since they are eternal.

Even if Empedocles is not completely right when he says,[253] without [adducing] a cause,

> But they reign in turn as times goes round,

so that the all is alternately at rest and in motion, nevertheless, insofar as what takes turns displays a kind of order of what is determined with regard to what is determined, he seems to speak in a more natural way. For since the parts of time are finite with regard to one another, there is a certain proportion (*logos*), but of the infinite to the infinite, of which Anaxagoras seems to make use, there is no proportion. However, he says, it is not enough for Empedocles and those who talk that way to speak of 'in turn', but they should have added the cause of this, and 'not merely postulate it nor consider it as irrational'.[254] But if the discussion is about some principle, they should confirm it through induction (for principles must be corroborated from what is posterior to them), but if it is about what comes after the principles, through demonstration. So it is here; for Love and Strife are not the principles and reasons for the fact that they reign and cause motion in turn and for equal [periods of] time. If they were, then 'being Love would be the same thing as being Strife',[255] but it is peculiar to Love to gather divided things into one, and to Strife to divide and to separate. The cause of the fact that they alternate and for the same [period of] time would be other, and those who speak this way should have given it beforehand, adding the other cases in which things are disposed and take place in this way. For the argument would be persuasive if it called upon what is evident as its 1186,1
witnesses, as the proposition that in the Whole, Love brings things

together, while Strife is separative and dissipative, seems to be credible from the fact that among us, too, Love gathers human beings together, while enmity separates them from one another; for since this is obviously credible in some cases, it seems to hold true of the whole as well. Likewise, credibility should also have been provided from something else for (the affirmation that) they predominate alternately and for equal times in the universe. But since someone could easily reply to what is said that just as love always brings things together, enmity separates them, and there is no need to search for the cause of this, so among us as well, combinations and separations always reign in turn, he says that 'it is not correct'[256] to assume the fact that something is always and everywhere disposed in a certain way as an indemonstrable principle. For many of the things that exist are everlasting and are everywhere disposed in this way, but they are neither principles nor indemonstrable, but have some causes and principles for being as they are. In mathematics, for instance, the fact that the three internal angles are equal to two right angles is everlasting and holds true in every triangle, but it is not a principle and an undemonstrable, but there is some principle of it, from which it is demonstrated. Likewise, the fact that summer and winter occur alternately is everlasting and everywhere thus, yet it is not a principle, but has principles and causes of such motion. Thus, he does not accept Democritus, who reduces the causes concerning nature to the principle that 'things also took place in this way previously'.[257] no longer thinking it well to seek the principle and cause of what always occurs similarly, and failing to understand that in the case of things whose everlastingness is a self-evident and indemonstrable principle, it is correct not to seek the cause of what is always that way, but in the case of things for which the everlasting is not a principle, one should give principles and causes. For not everything everlasting is a principle, even if a principle is necessarily also everlasting; for 'everlasting' has a greater extension than a 'principle'. Thus, when a geometer is asked to prove that [the internal angles] are equal to two right angles, he does not content himself with saying that this is everlasting, and a principle, and undemonstrable; but he adds a demonstration. Therefore, let Democritus not consider that the fact that what occurs always takes place in this way is a sufficient cause for the union and combination of atoms. It has often been stated that both Empedocles and Anaxagoras said the intelligible world is united, and the sensible world separated. But since both always exist, they assumed change in time to indicate their order, and the generation of the sensible from the intelligible, Empedocles is also taking it to indicate the conversion (*epistrophê*) of the sensible toward the intelligible.

Chapter 2

Translated by Michael Chase

252b7-27 That there was no time[, nor will there be, when there was no motion, or will not be, let this be said. The arguments contrary to these are not hard to refute. It might seem, to those who envisage things from this perspective, that it is above all possible for motion to exist at a given moment [after] not having existed at all, first because no change is eternal. For all change tends by nature to be from something and toward something, so that it is necessary that the limit of all change be the contraries in which it comes about, but nothing can be moved *ad infinitum*. Moreover, we see that a thing is capable of moving while not being moved or having any motion within itself, as is the case with inanimate things, which move at a given moment without either a part or the whole being in motion, but at rest. Yet if it does not come into being [after] not existing, it would have been fitting for it to either be always in motion or never. This is much more evident in the case of animate beings. For sometimes, without any motion being present in us, but being at rest, we nevertheless move at a given moment, and the principle of motion comes to be within us from ourselves, even if nothing moves us from without. For we do not see anything similar in the case of inanimate beings, but something else always moves them from without; yet we say that animals move themselves. Thus, if they are completely at rest at a given moment, motion would come to be in something unmoved from itself and not from without. But if this is possible in an animal, what is there to prevent it from happening in the All as well? For if it comes about in the small world, then it does so in the large one as well. And if [this happens] in the world, then [it happens] in the infinite], if it is possible for the infinite to be in motion and at rest as a whole.

Problems are always corroborated in three ways: through their principal demonstration, and through the refutation of those who state the 1187,1
contrary, and through the resolution of the objections against the problems. Therefore, having established through primary arguments that motion is everlasting, and having refuted those who seemed to say it is not everlasting, because they say things that are fictional and unnatural, he now moves on to the third point: refuting the objections that seem to shake his argument, and demonstrating them to be false. There are three objections, of which the first is as follows: no change is everlasting. And the demonstration of the argument is this: every change is from something, and into something; for change is into one thing after another, and from contrary into contrary. But change from contrary to contrary has the contraries as its limits and terms (*horous*).

A [change] that has its limits at infinity could not occur. Therefore, every change and motion could not take place to infinity, or in other words, no change that is one and continuous is everlasting; for 'every one is not' is equivalent to 'none is'.[258] But if no motion, as long as it is one, is everlasting, neither would motion in general be everlasting. But this objection does not seem to confute [the proposition that] motion always exists, for what prevents there from being another [change] prior to each change, which is finite, and another one after each one? Rather, this objection undertakes to confute [the proposition that] there is one continuous and everlasting motion.

The second objection against [the proposition that] motion is ungenerated is based on self-evidence. For we see many things that are not previously in motion, nor do they have any motion within them, either in whole or in part, changing into [a state of] being moved, as is seen 'in the case of inanimate beings'.[259] For stones and pieces of wood, which are not in motion either in whole or in part, but at rest until that point, are moved at a certain moment, and begin to be moved, changing from complete rest to [a state of] being moved. Yet if motion were everlasting, these things should either always be moved or never be moved, lest motion, changing from [a state of] non-existence, should be generated and no longer everlasting. As things are, however, these things both begin and once again cease to be moved. This objection too, however, which objects to the [proposition that] motion is ungenerated, and the one after it, merely prove that some motions occur after having not existed previously, but not, however, that there is no other motion before them, nor that there is not another one after them.

Third, and much more apparent 'in the case of ensouled beings' seems to be [the fact] that motion comes about after not having existed previously. For although inanimate things change from being previously at rest to being moved, what makes them move is itself moved from without, so that motion is not proved to be generated by inanimate things simply by the fact that another motion pre-exists the generated motion of these things. In the case of animals, however, it is not possible to say this, since they appear to be moved by themselves from complete rest, without anything moving them from without, which, since it is in motion, was able to cause motion. For then it seems that motion is evidently generated after not having existed, if it is moved from itself, not having been in motion previously. 'But if it is possible in an animal'[260] to begin a motion that did not exist previously (but an animal is a little world, set in order and administered in accordance with some order and nature), 'what prevents the same thing from occurring' in the whole world, viz. that it change from complete immobility to being moved from itself? Thus motion, which did not exist previously, comes into being. But if he says this is possible 'in the world', it is also possible 'in the infinite',[261] either that of Anaxagoras (as Alexander understood, since the mixture, arising from the uniform stuffs,[262] which are infinite, must be infinite), or that of Anaximenes, Anaximander, and in general those who postulate one element infinite in magnitude, or of both.[263] He also added 'if it is possible for the entire infinite to be moved or at rest', because it has been proved that what is infinite in size cannot move by the locomotion by which self-moved things carry out their motion. For it is not in a straight line, since there is nowhere toward which the infinite might progress, not in a circle, since in the infinite there is no up or down, or end and middle. Many other proofs about this were written in the third [book] of this treatise.[264]

> **252b28-253b2** The first-mentioned of these[, viz. that motion toward opposites is not always the same and one in number, is stated correctly. For it may be necessary, if the motion of one and the same thing cannot always be one and the same. I mean, for instance, whether of one string the sound is one and the same, or always different, as long as the string is similarly disposed and moved. Whichever the case may be, however, there is nothing to prevent it from being the same by virtue of being continuous and eternal.] Yet this will be more clear from later considerations.

Having set forth the three objections to [the proposition that] motion is ungenerated and everlasting, he replies to the first one, which says that there is no ungenerated and unfailing motion, since every motion is from contrary to contrary, and from limit to limit. Thus, if something always moved from contrary to contrary, this motion would not be one or uninterrupted, but always a different one would take place at different times, owing to the [state of] rest between them. Resolving this argument, then, he says that it is true that the motion that takes place toward opposites cannot be one and continuous. For even if the change took place again and again for the object in motion according to the contraries, it is impossible for motion taking place in this way to be one and continuous. Thus, he says 'it may be necessary'[265] that such motions be neither ungenerated nor everlasting, since it is not possible for such a motion of one and the same thing to be always one and the same. For if it is proved that it is impossible for the motion that takes place toward opposites to be one and continuous, which has not yet been proved, it would be proved that no motion taking place toward opposites is everlasting. But it will be proved that it is necessary that if the object in motion remains the same, the opposite motions must be divided by standing still (*stasis*). This is why until now he added 'perhaps', and shortly afterwards 'but nevertheless, however this may be',[266] since this has not yet been proved. Yet having said 'since it cannot always be one and the same',[267] he makes what has been said clear by an 1189,1
example. For instance, when a string is struck again and again, and passes over to the contraries, and is moved in a similar way, is the

sound one, continuous and the same, or always other? After all, if the sound, which is what we sense most of all, is one and the same, then the motion is also one and continuous; for the sound follows the motion, and if it is always other (as will be proved), the same will hold true for the motion. Having thus delimited the question in this way in the case of the string, 'however it may be', he says, in the case of such motions that take place from contrary to contrary, whether in their case motion is one or not one, 'there is nothing to prevent'[268] there being some other motion in addition to those that take place toward the contraries, which is continuous and one, and therefore everlasting, which he says will become clear from what will be said afterwards. He is talking about circular motion, for this is the [motion] he will prove is the only one that does not take place toward the contraries, like the rest, and is therefore one, continuous, and everlasting. Of the first argument, he says it is true that motions that take place toward the contraries are neither continuous nor everlasting, but it is not true that everything in motion is moved from a contrary and toward a contrary; for it will be proved that there is a motion that is one and continuous. See how, although he was able to resolve the objection on the basis that it did not confute the existence of motion, but the [proposition that] it is one and continuous (after all, this was enough to maintain unhindered the [proposition that] motion always exists), but referring to its concept, he proved that even if not every [motion] is continuous, there is nothing to prevent there being one that is continuous.

> **253a2-7** There is nothing absurd about being in motion while not being moved[, if the mover is sometimes external and sometimes not. But it should be investigated how this could happen, that is, so that the same thing, being under [the power of] the same mover, is sometimes in motion and sometimes not. For the person who says this does nothing other than wonder why it is not always the case] that some beings are at rest and others in motion.

Moving on to the solution of the second objection, which confutes the ungenerated [aspect of] motion from the fact that inanimate things, previously at rest, begin to be moved at a certain moment, and cease their motion at a certain moment, he also solves it by [the fact that] motion comes into being and perishes. He says that the fact that what is inanimate, although not previously in motion, is moved, is not probatory of the fact that motion is generated. For it is not proved, by the fact that these things begin and cease to be moved, that motion is universally generated, but that certain things, which are external and in motion, when they are present, are the causes of their motion. Thus, their motion is pre-existed by the motion of the things that move them,

and their [motion is pre-existed] by [the motion of] the things that move them, and so on. Having said this, he recommends that we enquire, since it is worthy of enquiry, how 'the same thing, movable by the same mover, is sometimes moved and sometimes not'.[269] For if the other things are moved by the things that are always in motion, how comes it that, while the things that cause motion are always in motion, the things they move not always in motion, but are sometimes in motion and sometimes not? For the person who raises this difficulty is not establishing that motion is generated, but he is seeking the starting 1190,1
point of what is now proposed for investigation,[270] that is, 'not everything in motion always moves, or everything at rest is always at rest', if at least the things moved by something else always have what will move them, but there are some things that are alternately at rest and in motion. These matters will be articulated shortly.

> **253a7-21** The third one would appear to raise the most problems[, viz. that motion comes to be internal without previously being so, which is what happens in animate beings. For while they are previously at rest, they afterwards walk, with nothing outside them apparently causing the motion. But this is false, for we always see some one of the natural parts moving in an animal. Yet the animal is not the cause of this motion, but perhaps the environment. We say that it does not move itself by every motion, but only by locomotion. Yet there is nothing to prevent, and perhaps it is even necessary, that many motions should come to be within the body from the environment, and of these, some are moved by discursive thought, others by desire, but the former already moves the entire animal, as occurs in dreams. For when no sense motion is present within it, but when some motion is [present], the animals awaken once again.] Yet it will be more clear with regard to these things, too, from the considerations that follow.

He solved the second objection, which derived from things that are inanimate and moved from without, on the basis of the fact that there is nothing absurd that they should sometimes be moved by the presence of what moves them, which is itself in motion, for he proved that another motion precedes the one taking place now, and sometimes is at rest, when the thing causing motion is not present. He now moves on to the third [objection] and sets it forth clearly, [viz.] that an ensouled being, which previously at rest and then walking, without anything external causing its motion, 'as it appears',[271] makes it plain that motion is generated. Having borne witness that this objection raises more difficulties than the others, he solves it, stating that it is false to assume that animals that were previously at rest are moved by themselves without anything external causing their motion; for, he says, 'we always see some one of the natural parts moving in an animal',[272] that is, one of the things that are naturally co-present to it. For it is either warmed, or cooled, or dried, or moistened, or it changes passively according to some other natural qualities. Yet perhaps what he calls 'natural parts' are what later philosophers also called 'natural capacities', attractive, retentive, alterative, and expulsive;[273] for these are always active in the bodies of ensouled beings, and especially of animals, and animals do not possess them in accordance with what is self-moved, for that motion is only local. But having said that 'the environment'[274] is the cause of the connatural motions, he added 'perhaps', both here and shortly afterwards, either because 'of the natural parts' was said without demonstration, or because not all of them derive from the environment. After all, they are in motion when they are nourished and grow, and again when they are diminished, and they are always in motion in these respects, neither simply through the agency of the environment, nor by the locomotion of animals. However, the motions in accordance with the passive qualities – warmths, as I said, and coolnesses, separations and combinations, and such like – they have primarily from the environment. 'There is therefore nothing to prevent', he says, 'or rather it is perhaps necessary'[275] that many of these natural motions come to be within the body from the motion of the environment, and that some of them sometimes set the discursive intellect in motion, and sometimes the appetite. For the discursive intellect, having taken cognizance of an affect through sense perception, deliberates about whether it should avoid it or submit to it, and it thus sets the animal in motion, by that locomotion that seems to derive from itself. Sometimes, however, when the apperception of something pleasant or painful occurs, appetite and avoidance are set in motion, as in the case of irrational animals, and these set the animal in motion locally, for instance toward drink or food, although it seems that this motion did not exist previously, but [the motions] in accordance with discursive thought and appetite preceded it, and before these the passive [motion] in the body, which the environment – itself in motion – sets in motion. He states the cause of locomotion more exactly in the *On the Movement of Animals*. He confirms [the proposition] that there is always some motion in animals, not deriving from themselves, but from the environment, which becomes the cause of locomotion for animals, by what occurs in dreams. For sleepers are not in motion by virtue of locomotion, nor do they have sense perception, which are the proper characteristics of animals *qua* animals. Yet they are nevertheless moved by many motions: those of respiration and the other connatural [motions], and those in accordance with digestion, and in addition representations and the passive [motions] deriving from the environment, when something within them is heated or cooled. But when the exhalations from food overcome the brain or the heart all at once, and are cooled down, they become the cause of sleep; when they are warmed, separated, and become finer, [they become the causes of] rising, as he proves in the *On Sleep*.[276] And again, animals have sense-perception and are in locomotion, deriving the beginning of this motion from some motions by which they were moved when asleep, so that if there were a complete lack of motion at that time, they would not have risen in the first place. Thus, in animals too, external motions are the causes of locomotion, which, since it [seemed] to begin from the animals themselves, first gave rise to the objection based on animals. Yet how these things are and come about, he says, 'will be clear from what follows',[277] for he will prove how animals are moved, and by what, and in what way self-moved things are moved in general.

It is worth noting in this passage whether the animal is not the principle of any locomotion, but all of them have their principle from without. For it is not only toward drink or food, or some such thing that we are moved, through the heat and dissipation in the body, but we also carry out many local motions for the sake of learning, profit, honour, power, and the fulfilment of duties; and not only we, but irrational animals also rush toward certain places when their [faculty of] representation is set in motion. Yet, one might say, the principle of the motion of these things too is also external; for the object of the desire, by moving the imagination or the discursive intellect, arouses the animal to locomotion, while the object of desire causes motion while remaining unmoved. In general, however, what does the person who has solved the objection based on animals wish to prove? That the locomotion of animals is preceded by another motion, whether of discursive thought or imagination, and this one [is preceded by] the motion in the body, and that one [is preceded by] the motion in the environment. And what does this one [seek to prove]? For such a motion, consisting as it does in different species – local, imaginative, and 1192,1
alterative – is not one and continuous. Nor, however, is such a motion everlasting. For an alteration of the body does not necessarily set the discursive intellect or the appetite in motion, nor does the latter, once it is set in motion, necessarily set the animal in locomotion, nor does the latter's locomotion necessarily set something else in motion. How, then, can the [proposition that] motion, being one and continuous, is ungenerated and everlasting, be corroborated on the basis of these things? Perhaps, then, his purpose in solving these objections is not to prove that all particular motions are one continuous motion, ungenerated and unfailing. After all, when solving the first objection he clearly granted that motion toward contraries is not numerically one, and he said that this is necessary, since the motion of one thing cannot always be one and the same. But since the objections against the beginninglessness of motion adduced some particular motions that occurred from some principle, I think he directed his solutions in reply to this point alone, proving that the motions proposed as being beginningless are not beginningless, but have some principles, but not, however, that they are

ordered to one beginningless [motion]. For it is not the [motion] that derives from them that he wishes to be unengendered and imperishable, but the one he indicates when he said that even if motions toward the contraries are not continuous and one, nevertheless nothing prevents there from being one that is the same, by virtue of being continuous and everlasting. But this is circular motion, toward which his entire argument about the one, beginningless, and endless motion tends, for it precedes all the motions that have a principle, and they exist through it, which is both before every particular motion and after every one, because they are reduced to it. But this will be more clear in what follows, when the one, everlasting motion is investigated. Meanwhile, he has so far proved that motion is unengendered and beginningless, not insofar as it is one and continuous, but because before every motion there is another one, and another one after each [motion]. Starting out from this point in what follows, however, he will discover the one, continuous, and everlasting circular motion, which is the cause of these unceasing motions as well. But why did he propose to investigate whether motion is everlasting or not, but not about rest? Because rest is the privation of motion, but no privation is everlasting, because it cannot be of an everlasting reality. Nor does privation co-exist with habitus, but it receives in succession motions that come into being and perish. Standing-still (*stasis*) and abiding (*monê*), however, since they are no longer privations, but the contraries of motion, are also beginningless and everlasting, as is motion in a circle; for it always takes place around something that remains.

Chapter 3

Translated by Michael Share

1193,1 **253a22-32** The starting-point of the investigation, which was also associated with (*peri*) the puzzle posed [earlier],[278] [will be why it is the case that some things (*onta*)[279] are at one time in motion and at another at rest again. Now, it is necessary that all things are either always at rest, or that all things are always in motion, or that some are in motion and some at rest; and again, in the case of these last, either that those that move are always in motion and those that rest [always] at rest, or that all things are by nature equally inclined to move and to be at rest, or that the only remaining (and third) [possibility applies]. For it is [also] possible that some things are always motionless, others always in motion, and that [still] others partake of both [these conditions]. And this [last] is what we must maintain; for it holds the solution to all [our] puzzles] and is the culmination (*telos*) of this study of ours.

He says that the starting-point for the more precise articulation of the foregoing[280] is the same as that associated with the earlier puzzle[281] which asked how it is that the same movable is sometimes moved and sometimes not moved by the same motive [agency]. (He said[282] [there] that 'someone who asks this is in effect asking why it is not the case that some things are always at rest and others always in motion', but there are some that are sometimes at rest and sometimes in motion.) So he is saying that that puzzle and the puzzle as to how animals move have this same starting-point.[283] And, further, both with regard to the original question[284] as to whether motion is everlasting and with regard to the arguments raising objections[285] to the everlastingness of motion, he says[286] that it will be useful to start from basics[287] and carry out a division to determine how things stand in relation to motion and absence of motion.

Starting from basics, then, he performs an exhaustive division which progresses through all existing things: For it is necessary either (1) that all things are always at rest or all things are always in motion (and for all things the division is one into contradictories, since all things are either in motion or not in motion),[288] 'or (2) that some are in motion and some at rest' (and this three ways: either (a) those that move always move and (b) those that are at rest are always at rest, or (c) all things both move and are at rest by turns), or ('the remaining, and third', [possibility], he says) (3) that 'some things are always motionless, some always in motion', some at times in motion, at times at rest.[289] He called this last 'third' in relation to the first (either all things are at rest or all things are in motion) and the second ('or some things are in motion, others at rest'), which he divided three ways.

He calls this alternative, then, third after these two. [It is the one] which he says he 'must maintain', that is to say, demonstrate and prove. For all the apparent puzzles[290] will be resolved once this has been clarified: (a) the everlastingness of motion will not be eliminated on the basis that all motion is from one point to another (*pothen poi*)[291] if it is shown that not all motion is such; (b) the fact that motion arises in the [previously] motionless is not proof that there is generation of motion [in general] if it is not everywhere the case that motion arises without having existed [previously];[292] (c) and the puzzle which asks why moving things do not always move and things at rest always remain at rest will also be resolved if it is found that there is always something that is unmoving and always something in motion and other things that move at one time and are at rest at another. And in this way the best conclusion (*telos*)[293] will be given to the treatise on motion,[294] which will have proceeded[295] demonstratively from its beginning by way of its middle parts (*ta mesa*) to its end (*telos*).

Alexander says that [that] 'end' is to discover something that is always and continuously moved by [something] everlasting and unmoving. But perhaps Aristotle has said that the same thing [will be] both a solution to the puzzles and the end of the treatise, [namely,] the establishing of that third [option, the one that states] that some things are motionless, others always in motion, others with a share in both [conditions]. 1194,1

It is also possible, I believe, to construct a threefold division of the relationship of things to motion and rest in the following manner: Given that there are these three [states], (1) being always in motion (2) being always at rest, and, third, (3) being at times in motion and at times at rest, [their] relationships will be divided into three in three [different] ways, either in such a way that all things are in the same condition, or so that they are divided two ways, or three ways, into beginnings, middles and ends.[296] After all, what further [possibility] could there be? And if all things are in the same condition, either all things are always at rest, or all things are always in motion, or all things are sometimes at rest and sometimes in motion by turns. And if things are divided two ways, either some are always at rest and others always in motion, or some are always at rest and others sometimes in motion and sometimes at rest, or, conversely, some are always in motion and others at times in motion and at times at rest; and so this section of the division is also divided three ways. The third [division] is the true one.[297] If things are divided three ways, it is necessary that some should always be at rest, some always in motion, some sometimes at rest and sometimes in motion. And over and above these nine sections there is, I believe, no other relationship of things to motion and rest, since it is necessary either to be in motion or to be at rest or to be at once[298] in motion and at rest.

He spoke more properly when he described things that are never in motion as 'motionless'[299] than when he described them as 'at rest'.[300] Rest, being the privation of motion, is an attribute of things whose nature it is to also sometimes move. And, moreover, the division thus becomes one into contradictories, dividing [things] into those which are always in motion and those that are never in motion.

> **253a32-b6** Now, [to maintain] that all things are at rest and to look for an argument for this [position], [putting aside [the evidence of] sense perception, is an instance of feebleness of the intellect and a calling into question of a whole and not [just] a part;[301] it is not only directed against the natural scientist but against virtually all the sciences and against all beliefs, since they all make use of motion. Further, as far as objections relating to principles are concerned, just as in reasoning in the mathematical sciences they are of no interest to the mathematician, and similarly with other [disciplines], in the same way nor are they to the physicist in relation to the present topic;][302] for it is a basic assumption (*hupothesis*) that nature is a principle of motion.

Having set out the division of things according to [their relationship to] motion and absence of motion in the manner described (*houtôs*), he goes through all of the unsatisfactory sections of the division, so carefully as to even repeat the argument, because this approach, involving as it does strict demonstration, both, as he says[303] himself, furnishes a solution to the puzzles and, by referring the lowest of physical things up to the highest and bringing these to a culmination with the transcendent principles of things, provides the best conclusion[304] to his treatise on physics.[305]

First, then, he scrutinises[306] the [section] that comes first in the division, the one that says that all things are at rest, and employs four concise arguments against it.

One is that to look for an argument that proves that all things are at rest, 'putting aside sense perception', which so clearly contradicts [such a position] and declares that many things are in motion, [constitutes] a weakness of the intellect which results in a person failing to understand that in such questions arguments are no more to be trusted than sense perception. [Such a person] also fails to understand what things should be called into question – that it should not be things that are perfectly obvious – or what criterion[307] one should employ in what circumstances, [in particular] that one should for the most part make use of sense perception where sensible things are concerned and make it the starting-point of one's demonstrations.

[The] second[308] [is] that one who eliminates motion seems to elimi- 1195,1
nate just a part of the attributes of natural things, motion,[309] but without realising it he is arguing about not [just] a part but the whole. For if nature has its being in being the principle of motion for natural things,[310] when motion is eliminated nature is eliminated. And if this is eliminated, natural things, whose being consists in their having a nature, are eliminated.

[The] third [is] that one who says that all things are at rest does not eliminate just nature and natural things and is bringing objections[311] not just against the natural scientist but against all the arts and practical sciences, that is to say, those whose operation depends on motion. (The theoretical science of anything does <not>[312] require physical motion, which is why he adds 'virtually' when he says 'against all the sciences'.) And he is at odds with all the views of the natural scientists, indeed of all human beings, since they all make use of motion. And in view of this [passage] too it is, I think, important to note that Aristotle is not of the opinion that even those around Parmenides and Melissus[313] believe that all being is motionless, only that true, or intelligible, being is.[314]

Fourth he adduces the argument that just as it is not the role of the mathematician to argue against those who would eliminate the first principles of mathematics, or of the doctor against those who deny medical first principles, or of [the practitioner of any other science or art] against those who would eliminate his first principles, in the same way it is not the business of the natural scientist to demolish the present proposal that all things are at rest, for that nature is the principle of motion is assumed as a principle by natural scientists. [Aristotle] made the same point[315] against those appearing[316] to claim that being is motionless in the first book of the present treatise.

It is customary to call first principles basic assumptions (*hupothesis*)[317] because they are posited without demonstration and are assumed for the demonstrations that are [constructed] from them. In fact, one who demonstrates his first principles confesses that they are not [truly] first principles, for demonstrations [must derive] from prior [assumptions].

253b6-254a3 Roughly speaking, while it is also false to say that all things are in motion, [it is less counter to method[318] than this [position].[319] Nature was indeed established[320] as a principle of rest as well as of motion in the *Physics*,[321] but motion is nevertheless the natural [state]. And there are people who say that it is not the case that some things move while others do not, but that all things do, and all the time, but that this escapes our perception. Although they do not specify what kind of motion they mean, or [whether it is] all kinds, it is not difficult to reply to them. After all, it is not possible to increase or decrease continuously, but there is also an intermediate [state]. The argument is like the one about the dripping [of water] wearing away stones, or [plants] growing from them splitting them. For if the dripping has forced out or removed so much [stone] it is not also the case that [it] earlier [removed] half [the amount] in half the time. As with hauling a ship, so many drops set so much in motion, but a fraction (*meros*) of them [will] not [set] as much [in motion] over any [period of] time. The [material] removed is, it is true, divided[322] into further [parts], but none of these was removed separately but [all of them] together. So it is clear that it is not necessary that something should always be leaving [the rock] because the detritus (*phthisis*)[323] is divided to infinity, but the whole [amount] leaves [the rock] at one point in time (*pote*). The case is similar with alteration of any kind; if the thing that undergoes alteration is divisible to infinity, the [process of] alteration is not on that account also [divisible to infinity] but often takes place all at once, like freezing. Also, when anything[324] sickens, there must come a time during which it will recover and it [can]not change [its condition] at the limit of a [period of] time. And it must change to [a state of] health and not to anything else.[325] And so to say [things] alter continuously is too much at odds with the evident [facts]. After all, alteration is to an opposite [condition] but the stone becomes neither harder nor softer. And in the case of locomotion it would be remarkable if it could escape our notice[326] whether a stone was falling or at rest on the ground. Also, earth and each of the other [elements] necessarily remain in their proper places and are moved from them [only] by force. So if some [quantities] of them are in their proper places, it must also be the case that not all things are moving locally.] From these and other

> such [considerations], then, one may be confident that it is impossible either for all things always to be in motion or for all things always to be at rest.[327]

He turns to the second section of the first division, the one that states that all things are in motion, and says that this is in a way almost as false as saying that everything is at rest. In this case too perception clearly contradicts [the proposition], since there are things that are at rest. However, the [proposal that] all things are in motion goes, he says, is 'less' counter to the method of physics than the [proposal that] all things are at rest. This is because it is less destructive of nature. For even though, to the extent that nature was posited as the principle of rest as well as the principle of motion in the *Physics*,[328] one who eliminates rest also eliminates nature, to the extent that motion is more pertinent (*oikeios*) to nature than is rest, what such people say goes less counter to the method of physics. And [it is also so] because the former people eliminate not only motion but rest as well (for things that are at 1196,1
rest also have it in their nature to be in motion), while these people only eliminate rest. And further, since there are some things in nature that are never at rest but there is nothing in nature that is entirely without motion,[329] it therefore goes less counter to the method of physics to say that all things are in motion than that all things are at rest.[330] And it is also an indication that the view that holds that all things are in motion is not entirely foreign to the physicists that some of them are also of the opinion that all things (not just some of them and others not) are in motion, but that it does not seem so to us because it escapes our perception. The Heraclitans said that all things are in motion and Plato records their doctrine to that effect in the *Cratylus*. And Alexander says that, for those who posit them, the atoms, by being always in motion, also become the causes of the compounds that are formed from them, though not at a perceptible level, 'but even according to these people', he says 'the void is without motion'.[331]

Well, he says, to these people who say that all things are in motion, 'although they do not define what kind of motion they claim that all things display, or indeed [whether they mean that] all things [display] every kind, it is not difficult to respond'[332] by distinguishing the types (*eidos*) of motion and showing that it is not possible for all things to be always in motion in respect of any of them. Accordingly, he first brings against them a charge of talking vaguely and not making it clear with what kind of motion it is that all things are always in motion. Then he shows for each of the three types of motion that it is not possible for all things to always be in motion with respect to any one of them, and if not with respect to any one of them, then clearly not with respect to all of them either.

First he shows that it is not possible for a thing to grow continuously, nor, conversely, to decrease continuously, nor to move directly from growth to decrease,[333] but that there is immobility (*stasis*) and rest in between for [anything] that moves with each of these motions, because a thing that is growing does not switch over to getting smaller nor a thing that is getting smaller to growing but the former to increasing [in size], the latter to diminishing [in size]. So, when it becomes one of the things into which it was changing, it is clear that it comes to a halt and only then switches over to getting smaller. He will show that this is the case in what follows.

After saying that some claim that all things are in motion 'but this escapes our perception', and after instancing growth and decrease on the basis that they say the same of them, he first asks that theory (*doxa*) this account (*to legomenon*) of theirs resembles and, by showing that it is false, at the same time demolishes their account. For to claim, he says, that all things are always in motion but that this escapes our notice because it is only little by little (*kat' oligon*) is like the well-known story (*logos*) about the constant dripping[334] of water hollowing out the rock – 'a drop[335] of water, by persistence, hollows out a rock',[336] says Choerilus – the rock supposedly being worn away[337] by each drop but the gradual (*kat' oligon*) wear[338] being invisible to us. And similarly, [they say,] stone is at some point pulled apart and cracked open by a fig-tree, say, growing in a fissure in the stone, supposedly being pulled apart little by little throughout the preceding period as well even though [this] escapes our perception thanks to the gradual (*kat' oligon*) nature [of the process]. Just, then, as those who tell these stories claim that the rock that is worn away and the stone that is broken apart are in continuous motion throughout the preceding period but this escapes our perception because of its gradualness (*kat' oligon*), in the same way those who claim that all things are always in motion always say that their motion is continuous but that it escapes us because it is gradual – and [that] this is how it is with things that are growing and decreasing too.

So, having thus adduced a similar theory, he shows that it is false and that the wearing away of the rock and the splitting apart of the stone are not continuous. Even if the fig-tree has separated the stone a span[339] over a year, the seedling did not separate [it] by any part [of that] during the first day; and even if ten thousand drops have hollowed out the rock, just one drop did not hollow out any part [of it] at all. Instead, these [processes] are, he says, like hauling a ship. In that case too a hundred men at once can launch a ship but any one of them on his own[340] would not even rock [it]. And in the same way, although a given number of drops have removed a given amount of the rock, a single drop would not remove any part [of that] at all. For, even though the amount [of stone] removed from the rock might be divisible, it is not necessary that each drop should remove an amount (*meros*) corresponding to the proportion it represents of all [the drops]. After all, the whole [amount of stone] was removed by all [the drops] together. And just as in the case of hauling ships, a ship that has not so far been moved by fifty men has sometimes been set in motion (*saleuein*) when one man joins them, and in the case of a balance [a weight of], say, ten talents has not produced a tilt but an extra dram has depressed it,[341] in the same way a given number of drops has not had the power to hollow out the rock, but with the addition of one more, has been able to. And so in the case of things that grow or diminish as well, even though what comes in or goes out may be divisible to infinity, there is no necessity that there should without intermission always be something being separately added or always [something] being [separately] taken away,[342] but sometimes the whole of something [may be] added or the whole [of something] taken away. For it is not the case that because all the water has hollowed out so much [stone], half of it will hollow out half [that amount], or, more generally, that a given amount of water will remove a given amount of what is removed. The parts are not even present actually in what is being removed to allow each to be moved separately, but the whole is moved at once.

By 'detritus' (*phthisis*)[343] he means [the material] that leaves [the rock and] owing to which the damage (*phthora*) occurs.

It is clear, then, that if one thing is affected by another, it will not be affected by each and every (*to tukhon*) small part of it, and that, if one thing is affected by another during a given [period] of time, it is not also affected during every small part of that [period of] time but all at once during some one [of them].[344]

At this point we should consider whether he is here alluding to the well-known argument which is referred to as the sorites among the sophists in the form in which it is asked[345] in relation to the falling drops of water.[346] If, they say, the first [drop] has done nothing towards hollowing out the rock, nor has the second; and if that has not, nor has the third; and so nor has the last. How, then, can it be that 'a drop of water, by persistence, hollows out a rock'? So is he referring to this [argument] when he says 'the argument is like the one about the dripping (*stalagmos*) [of water] wearing away [stones]', or to another asked with regard to a fall of water [that takes place] all at once, which goes like this: If a given quantity of water falling at the same time has removed a given amount of the rock, did half of it remove half of that amount and a single drop a proportionate amount?[347]

Alexander, at any rate, evidently understood the argument in the latter manner. 'The theory',[348] he says, 'which states that, since the rock has been hollowed out by a given quantity of water raining down on the rock (perhaps he calls this [raining down] a "drop"[349] (*stalagmos*) because it drops down (*katastazein*)[350]), it was also hollowed out by each portion of the [water] that fell on [it] but the corresponding (*kata tosouton*) motion and wearing away (*meiôsis*) of [the rock] was invisible to us.'

Alexander appears to have followed the model based on the hauling of a ship, in which all the ship-haulers [worked] together, and [consequently] understood 'drop' as referring to all of the water at once, since[351] it is, I think, clear that 'drop' (*stalagmos*) is properly said of a [single] drop (*rhanis*)[352] and Aristotle's phrase 'a given number of drops' also indicates [this], since the drops are [evidently] separated and numbered. And when Aristotle said with reference to the portions [of stone] removed from the rock 'but none of them was moved separately but [only] together', Alexander took this as also applying to the water that moves [them] and said 'for the parts of the moving [agency][353] were not even present actually in the whole so as to be able to cause motion individually', in the understanding that it is all the water together that has fallen and removed some of the rock.

Perhaps, however the question was put,[354] it has to do with the same thing,[355] [namely, how motion can] escape perception by [being] gradual, whether the division into gradual [increments] takes place at the same moment, as in the case of the hauling of a ship or the removal [of material] by all the water at once, or[356] at one moment after the other, as in the case of the drops that hollow out the rock over a long period by [their] persistence. I do, however, think that the latter model fits the context better because the motion too, by not occurring all at once, escapes perception by being distributed. After all, neither growth and decrease, nor alteration, nor change of place [occur] all at once.

What, then, is being said is that while what leaves during diminution (*phthisis*)[357] and what enters during growth are, being magnitudes, divisible to infinity, there is nonetheless no necessity that the departure and the addition should involve the smallest possible [divisions of the material]; on the contrary, [such outgoings and incomings] sometimes leave all at once or are added all at once. Therefore these things[358] do not, by [being] gradual, decrease or grow to infinity unobserved by us, as they claim. If, as [Aristotle] says, 'there is something[359] in the middle', in which growing things stop [growing] when they have attained [their] due size, there is no necessity that they should grow to infinity or decrease to infinity. For, even if they continue to be nourished at that mid-point, being nourished is one thing, and growing another. And it is clear that even if some part of the rock is removed all at once, each drop has certainly done something towards making the rock weaker; and each of the ship-haulers has contributed his own strength to the total; and the splitting of the rock does indeed at times take place gradually without being observed, but at times it is split apart all at once, although [even then] the gradual growth of the fig-tree paves the way for the splitting by weakening the rock;[360] and although up to a certain point they proceed gradually without being observed, both growth and decrease come to a complete (*athroon*) halt once they come to consummation[361] upon reaching the [goal] towards which they have been striving.[362]

Aristotle has shown in the case of growth and decrease that, since the addition and removal [of material] is not [always] continuous but also takes place all at once, growth and decrease do not take place to infinity [and] escape [our] notice by [being] gradual, [and] he [now] states that the same argument applies to alteration of all kinds. For, [he says,] even if the thing that alters is a magnitude that is divisible to infinity (this is what 'divisible into an indefinite [number of portions]' means), its alteration does not also on that account take place one part after the other but all of it alters at the same time, and the alteration comes about all at once, like the curdling of milk and the freezing[363] of water and the like, [all of] which, though divisible, do not set one small portion after another[364] but change and congeal all at once.

'He has shown', says Alexander,[365] 'by what he says now, what he meant when, speaking against Melissus in the first book of this work,[366] he said "as though change did not take place all at once". By "all at once" he did not mean instantaneously (that would be false), but that the embarkation (*arkhe*) upon change is of the thing that changes as a whole and not of one part of it after another.'[367]

He shows that a thing that alters cannot be continuously in motion to infinity, as those who say that all things are always in motion claim, from [the considerations that] for things undergoing alteration the change is from opposite to opposite, and [that] their change takes place in time, and [that] there is a defined [state] to which they change and when they get there they stop [changing]. That is as far as change extends; for instance, anything[368] that is sick and is changing from sickness is, insofar as it is changing from sickness,[369] changing to nothing other than health (for that is the opposite of sickness). And the change in it certainly occurs over [a period of time] and not at the end (*en perati*) [of a period] of time, because every motion [takes place] during time and the end of the motion [occurs] at the end [of that period] of time. So, when it becomes healthy after being sick, it remains (*histanai*) in health. Indeed, there is no other [state] for something sick to move into, because, as we saw, sickness has no other opposite. So, if it does not have anything else to move into, 'to say[370] that [it] alters[371] continuously is too much at variance with the evident facts'. After all, towards what further [goal] will anything move after it has gained [the objective] that was the reason for its moving [in the first place]?

By saying 'there must be a [period of] time during which [anything] gets well and it [will] not [happen] at the limit of a [period of] time' he has shown that the change from sickness to health is a kind of motion; it does not take place instantaneously or at the limit of a [period of] time, but nor on that account does it go on to infinity.

This argument also applies in the case of the other types of change, since they all proceed (*ginesthai*) from opposite to opposite and strive, as a kind of goal, for [one] opposite, after which they are actually 1200,1
named[372] – generation from being, growth from increase, growing white from white, and upwards motion from 'up'.

That the motion involved in alteration is not continuous he also shows from manifest fact. For over a long period 'stone grows neither harder nor softer', but, as Eudemus[373] says, 'nor does anything emanate from it each day, since over many years the efflux would be greater than the body from which it flows, and yet', he says, 'many things, even small ones, are seen to endure, remaining untouched for a very long time'.

After mentioning a stone as a case of something that does not undergo alteration continuously, he adds that in the case of moving (that is, moving with respect to place) motion is likewise not continuous or uninterrupted in the case of a stone. For even though instances of growth and alteration may have escaped us, we certainly do not fail to observe whether a stone is falling or resting on the ground.[374]

And, speaking generally, earth and absolutely everything sub-lunar (for he is showing for all things in common that they do not exhibit continuous local motion) – if ever any of them are seen 'in their proper places', it is necessarily the case that they will not at that time be moving locally. He shows this as follows: All natural bodies remain [stationary when] 'in their proper places' and they move naturally towards those proper places and contrary to nature and by force away from them. Therefore whenever any of them are in their proper places, it is impossible for them to move locally, since if it is necessarily the case that anything that moves moves either by force or naturally and these things do not move locally in either of these ways, things that are in their proper places will not move locally at all.

The first two sections of the division have now been dealt with. He has established that all things are not always at rest by way of [an appeal to] self-evident fact, and he has established in addition that it is not the case that all things are always in motion.

254a3-15 Nor again can some things be always at rest [and some always in motion and nothing be sometimes at rest and sometimes in motion. As with the [views] described earlier, so with these, one must declare that this is impossible (for we see the stated changes occurring in the same things), and, in addition, that a proponent [of such views] is in conflict with the obvious [facts]. For there will be neither increase nor forced motion unless a thing previously at rest can be set in motion[375] contrary to nature. Accordingly, this account does away with generation and perishing. Further, virtually everyone considers moving a kind of coming to be or perishing. [A thing] becomes, or comes to be in, what it changes to, and what it changes from ceases being, or [it] ceases being in it[376] (*phtheiresthai enteuthen*). And so it is clear that] some things are in motion and some things are at rest [only] some of the time.[377]

Coming to the second section of the division, the one that states that some things are in motion and others at rest,[378] in this case too[379] he examines its first subdivision[380] (*diairetikon*), the one that states that the things that are in motion are always in motion and the things that are at rest always at rest, and shows that for things to be so disposed that some are always in motion and some always at rest and it is not a characteristic of anything to be sometimes in motion and sometimes at rest is both impossible, and impossible in the same way as the earlier proposals (*rêthenta*), because it is also self-evident that we see the same things at one time switch from motion to rest, at another from rest to motion.[381] (These are the 'stated changes' due to which some things are at one time at rest, at another in motion.) 1201,1

Secondly, he adds to [this] argument that someone who argues for this position is in conflict with the plain facts and adds in what respect this is so. Unless, [he says,] it is possible to be at rest at one time and in motion at another, 'there will be neither growth nor forced motion'. Indeed, generation and perishing will also be eliminated, and, as we shall learn, virtually all motion, whether natural or contrary to nature. So, if it is plain that growth and forced motion and [all] the other kinds of change exist, it is clear that, by doing away with [the possibility] that some things are at times at rest and at times in motion, [such a person] is in conflict with the plain facts.

That there will not [indeed] be forced motion he indicates by [writing] 'unless a thing previously at rest can be set in motion contrary to nature'; for something moved by force is moved contrary to nature, and something moved contrary to nature leaves the place that is proper and natural for it, [the one] in which it is its nature to rest; so if it is not moved after previously being at rest, there would not be unnatural and forced motion.

He seems to leave aside the question of how growth is eliminated as a result of a position of this kind. However, Alexander well says[382] that this too is shown [to be so] by the same [argument], for there will not be growth 'unless a thing previously at rest can be moved contrary to nature'. For example, nourishment which was previously at rest is transformed (*alloiousthai*) and carried to the higher parts and the sides [of an organism – parts] to which it is not its nature to travel naturally – and is [there] assimilated and absorbed, and when there is more of it than [the sum of] what has left [the organism], it causes [the organism] that has taken it in to grow.

One can also show that unless resting after motion is possible growing is not possible from what was said previously. For something that is growing neither grows continuously nor moves directly from growth to decrease, but once it reaches the goal (*horos*) the growth was striving after, it stops [growing] and [only] then switches over to decrease. And so the [idea that] everything that moves is always in motion is ruled out by [the phenomenon of] growth and the [idea that] everything that is at rest is always at rest by [the phenomenon of] forced motion.

Not only growth and forced motion are eliminated unless there are some things that can be at rest and in motion by turns, but generation and perishing are, as he says, also eliminated by this account. For just as something that is growing exhibits at that time the motion associated with growth – which it did not exhibit before [it began] growing – so too [*mutatis mutandis*] does something that comes to be or perishes. If something that is growing exhibited the same motion before [it began] growing as it does while growing, either it is not even now exhibiting the motion associated with growth, so that that nothing that is growing [actually] grows, or it was also growing earlier when it was not yet growing. And things that come to be or perish likewise have a beginning to [their] generation, and, once they have come to be, an end [to it], and similarly with perishing. And so, if things that move must always be in motion and things at rest must always be at rest, there will be neither growth nor decrease and neither generation nor perishing. After all, the changes involved in these [processes] both begin from the [state of] rest that is associated with them and [also] come to an end. And, further, a thing that is not able to come to be does not come to be in the first place, and a thing whose generation has no limit and which cannot come to maturity cannot come to be; for a thing that has come to be stops [coming to be] and ceases the motion associated with that [process].

1202,1 And, moreover, the account that eliminates unnatural motion eliminates generation and perishing, for the generation of some things is the perishing of others,[383] and it is not while remaining in a natural state that each thing perishes but while moving contrary to nature. And in something that is coming to be nothing occupies its proper place either, and for that reason things that come to be are also done away with.

It may be that the [words] 'accordingly, this account does away with generation and perishing' are equivalent to 'this account, which states [that nothing] changes from rest to motion or from motion to rest, also does away with generation and perishing'. For everything that comes to be will come to be to infinity, and similarly [with everything] that perishes, and that being so it would be impossible for anything to come to be or to perish.

Without going into every kind of motion or change, if the theory under discussion eliminates generation and perishing, it would also eliminate all motion. For moving, broadly understood, seems to virtually everyone to be coming to be and perishing – ['virtually'] because it would not, of course, have been accurate to say that it seems so to [literally] everyone. The reason for its seeming [so] to virtually everyone – or for moving being virtually a kind of coming to be or perishing, since 'virtually' can be construed in this way too – he supplies for all motion in common. For, since every motion and every change is from something to something else (for it is of one thing after another), '[a thing] becomes what it changes to', either, in respect of substance, as in the case of a human being, or, in respect of quantity, as in the case of larger or smaller, or, in respect of quality, as in the case of[384] white or hot); while if the change is one of place, the thing that changes does not become what it changes to, that is, up or down, or, to put it generally, a place, but comes to be in it. And 'what it changes from' – a substance, or larger, or white – 'ceases being'; for [instances of] human being or larger or white cease to be when they change to the opposites of these. And if [the motion is] from a place, the thing that leaves it ceases being in it.

He was right, then, to say that an account that eliminates generation and perishing also eliminates all motion. [Aristotle] himself refrains from arguing this through because it is obvious, but he does draw this conclusion from the refutation of the [proposition that] some things are always in motion (as long as they exist, of course) and others always at rest when he writes: 'and so it is clearly the case that some things are at times in motion and some things at times at rest'. (He means that there are some things that are in motion or at rest by turns and not always [in] just one [of these states].) After all, it is clear (1) that things that are at times in motion are at times at rest and (2) that things that are at times at rest are at times in motion and (3) that one should not think of things that are at times in motion as always in motion or of things that are at times at rest as always at rest, but of both as sometimes in motion, sometimes at rest.

> **254a15-22** The assertion that all things are sometimes at rest and sometimes in motion [must now be confronted with our recent arguments. We must, now that the present distinctions have been made, start again from the same point we started from previously.[385] Obviously either all things are at rest, or all things are in motion, or some things are at rest and some in motion. And if some things are at rest and others in motion, it must either be that all things are sometimes at rest and sometimes in motion[386] or that some of them are always at rest and others always in motion] and [yet] others sometimes at rest and sometimes in motion.

Having divided[387] the section that states that some things are in motion and others at rest three ways into (1) [the case that] the things that move are always in motion and the things that are at rest always at rest and (2) [the case that] all things are sometimes in motion and sometimes at rest and, third, (3) [the case that] some things are always 1203,1 motionless, others always in motion and [yet] others with a share in both [these conditions], and having shown[388] that the first section, the one that says that the things that move are always in motion and the things that are at rest always at rest and that nothing switches from being at rest to being in motion, is incorrect, and being on the point of turning to the second section of the present division (the one that states[389] that all things are sometimes in motion and sometimes at rest) and bringing it forward for refutation, he returns again to the original division, both reminding us of the earlier sections of the division and adding something to what was earlier said against them.

After the distinctions that have just been made, he says, 'we must start again from the same point we started from previously', and the same division must be considered again – which [division] he sets out clearly but lacking one section, the one that has just been shown to be incorrect, and which is perhaps omitted now for that reason. (This [section] is the one that stated that some things are always in motion, some always at rest, and that there are none that are sometimes in motion and sometimes at rest.) It also could be that it is left out now because it is just as impossible as [those that state that] all things are always in motion or [that] all things are always at rest and is ruled out in the same way as them, as he will tell us later.[390]

> **254a22-b4** Well, that it is not possible for all things to be at rest [was stated earlier, but let us repeat it now. For even if it really is the case that things are as some people say being is, [namely,] infinite and motionless, it by no means appears [so] to perception, but [on the contrary] many things [appear] to be in motion. Since, then, false opinion, or opinion of any kind, exists, motion also exists – and likewise if imagination [exists], or if [anything] sometimes seems this way, sometimes that; for imagination and opinion are thought to be kinds of motion. But, on the one hand, to investigate this question [at all], and to seek an argument in cases where we are better off than to need an argument, is to discriminate poorly between the better and the worse, what should be taken on trust and what should not be taken on trust, [what is] a principle[391] and [what] not a principle. And, on the other hand, it is just as impossible for all things to be in motion or for some to be always in motion and some always at rest. A single proof does duty against all these [theories]: we see some things that are at times in motion and at times at rest. And so it is similarly impossible for all things to be at rest or all continually in motion] and for some to be always in motion and others always at rest.

Earlier, replying to those who say that all things are at rest, he said[392] that it was feebleness of intellect to dismiss the proper criterion of sensible things, sense perception, and attempt to eliminate what is manifest to perception by means of arguments, and that someone who would eliminate motion is not arguing, as it might seem, about one aspect of natural [phenomena], [namely] motion, but, since nature is the principle of motion, about the whole fabric (*sustasis*) of nature; and that he eliminates not only nature and natural [phenomena], but all the arts and sciences, the former of which depend on motion in their activities and the latter of which are agreed that motion exists and appeal to motion when explaining the causes of all things that come to be; and, finally, that the natural scientist cannot even converse with someone who would eliminate motion, because [such a person] is eliminating the [fundamental] principle of natural [phenomena]. Now, taking up once more his argument with these people, he adds that even for those who would eliminate motion it follows that motion exists; for even if, [he says,] what they say is true, [namely,] that being is infinite and motionless (as he reported in the first [book] of the present work Melissus holds), it certainly does not appear to be so as far as perception is concerned, but [on the contrary] many things seem to be in motion. And, if false opinion and imagination exist, it is clear that motion also exists; for imagination is the persistence[393] of actual perception and opinion is assent (*sunkatathesis*)[394] to imagination and perception is 1204,1 motion and [takes place] through motion.

'After saying "anyway, if false opinion exists", he added "or opinion of any kind", because false opinion and false imagination are not motions *qua* being false, since all imagination and opinion is motion. And so even people who talk this way, even if they were telling the truth and setting out the facts as they appeared to them and as they believed them to be, would [themselves] be in motion. So, from the very words they used in seeking to eliminate motion, it would follow that for them there is motion.'

It was in this fashion, and in these very words, that Alexander expounded the present text. I do, however, believe that it is worth spending time over the question of how, if it really is the case that being is motionless, it does not seem so to perception, but many things are in motion.[395] How is it that true perception contradicts the truth? And, further, if things are really as they say, how is it that opinion is [therefore] false? (For the words 'since, then, false opinion exists' seem to be stated as consequential upon what precedes.)

Well, perhaps he concedes[396] to those around[397] Melissus that they are right when they say that being is motionless, and, assuming that the motion that is apparent to perception is falsely opined[398] [to be real] as a consequence of its being true that being does not move (for if this is true, motion is false), reverses[399] their argument by concluding that if it is true that being is motionless, motion exists. The structure (*sunagôgê*) of the argument is, I think, something like this: If it is true that being is motionless, there is false opinion on the part of those who believe that many things are obviously in motion as perception suggests (*kata tên aisthêsin*).[400] But if there is false opinion, then clearly for them[401] too non-being exists. And if non-being exists, the void exists. And if this exists, motion exists. This argument does indeed run in the opposite direction to the one advanced by them. For since they claim non-being is not in [the realm of] being, the void could not exist, and if the void does not exist, motion does not exist, for motion must be either through a plenum or through a void, and through a plenum is impossible.

So if there is false opinion, there is non-being. And if there is non-being, the void[402] is not eliminated; and if the void exists, motion is not eliminated.

[It was my conviction] that Aristotle would not have written 'since, then, false opinion exists' and then added 'or opinion of any kind' without a reason that led me to this idea for the interpretation [of the present passage].

But how is it that if opinion and imagination exist, natural or bodily motion – the subject of our present enquiry – also exists? After all, according to Aristotle they are not motions.[403] Alexander has a good answer to this objection. He says that perception is a motion, that imagination is the persistence of actual perception, and that opinion is assent to imagination. So, if opinion exists, imagination and perceptual activity exist, and if that exists, so does natural motion. 'And if', he says, 'people sometimes think one way, sometimes another, on some issues (and indeed Aristotle himself believes that opinions about things in human beings[404] are constantly changing[405]), that too means motion exists, because it is universally the case that opining does not occur
1205,1 without motion, since it follows upon imagination and that upon perceptual motion'.

Next Alexander formulates the argument (well, I think) as follows: 'If nothing there is moves and there is no motion, but some things [nevertheless] appear to move, [then] imagination exists; but if imagination exists, [then] motion exists and some things move; so if nothing there is moves and motion does not exist, [then] motion exists.' [This] reversal of [their] argument makes use of the same materials (*keimena*)[406] except that Alexander has used imagination instead of false opinion.

And one should [moreover] observe that Aristotle clearly states here that 'imagination and opinion seem to be motions' and not that they follow upon motions, unless perhaps the addition of 'seem' indicates that the opinion of others is being given.[407]

After making the above response to those who say that everything is at rest, he introduces [once more] some of what he said earlier with some additional detail. [This time] he teaches us about the different kinds of knowledge. [He tells us] that one kind is self-warranting, or without [explicit] warrant,[408] and does not need to be warranted by means of an explanatory account (such are the so-called immediate propositions, which are derived from common conceptions),[409] the other a type of knowledge that acquires warrant through the medium of an explanatory account (of such a kind is the middle term in syllogisms). And it is clear that self-warranting and self-evident knowledge is superior to the kind that has need of an explanatory account. In the case of the self-evident kind we have knowledge already and are better situated than when there is need of an account and an explanation; in the case of the other kind we are still searching and need some proof. So, when we seek explanations and accounts of things that are self-warranting we are, out of ignorance of what is better and what worse and what self-warranting and what not such in [these] kinds of knowledge, endeavouring to exchange the better [kind of] knowledge for the worse.

They say, for instance, that when someone, as an exercise, was proving, in the style of the argument of Zeno that uses infinite dichotomy, that motion could not exist, Diogenes, who was present, stood up and walked away, offering exactly the same advice as Aristotle does here, [namely,] that in matters where we have self-warranting knowledge there is no need of logical subtlety. The kind of person [who thinks there is] is not even aware, he said, of the criteria appropriate to [different] realities (*pragmata*). Just as someone who puts forward sense perception as the criterion in the case of knowledge of the intelligibles is all at sea and misses the mark, so too would someone who spurned sense perception as the criterion in the case of sensible things and looked for an [explanatory] account be [likewise] all at sea.

Moreover, such a person, if he indeed questions a thing that has warrant,[410] does not know 'what is to be taken on trust and what not to be taken on trust' and he will on that basis go on to question whatever is discovered and given warrant through his questioning, if for him what is warranted is not immune from questioning.

And anyone who questions the self-evident does not even know the difference between a principle and what is not a principle, since principles differ from the things that come after them [precisely] in this, in being self-warranting and not needing investigation.

He says that both those who claim that all things are in motion and those who say that some things are always in motion and others always at rest are making claims that are just as impossible as [the claim of] those who say that all things are at rest; and he adds as the reason for the similarity [of their claims] that a single warrant is sufficient in all these cases, [namely,] that some things are seen to be at one time in 1206,1
motion and at another at rest. Since this is self-evident, all these [positions] are proved impossible.

By writing 'or some things are always in motion, others always at rest' he has also included in the refutation the things[411] he omitted in the immediately preceding division. Perhaps he omitted them then because they could be refuted in the same way as the others[412] – and by way of the very same self-evidence on the basis of which he[413] also states that [their] similarity is obvious.

254b4-6 It remains, then, to consider whether all things are such [that they [both] move and remain at rest, or whether some things are such while others are always at rest and [yet] others always in motion;] for it is this last that we must show to be the case.

Now that it has been taken as self-evident that some things are manifestly sometimes in motion and sometimes at rest and the rest of the sections of the division have been refuted by the use of this [observation], [only] two still remain in which this clearly observed [phenomenon] is [present]; for either all things are such that they are sometimes in motion and sometimes at rest, or that third section of the entire division is true, the one which states that (1) some things are (as was just stated) so disposed as to be sometimes at rest and sometimes in motion (2) others so as to be always at rest and (3) others so as to be always in motion. This was the third section of the original division and he proposes to show that this is the true one. For, once this one has been shown to be true, both all those that were refuted earlier and the one proposed along with this one (the one that reads 'either all things are such that they are sometimes at rest and sometimes in motion') will, since they cannot co-exist with the true one, be rejected as false. And once this has been shown [to be true], it will also be shown that motion is not generated and the difficulties raised in relation to the ungenerated [nature] of motion will be resolved.

One should note that it would have been possible for two different (*allos*) sections (which would divide things in two, some of them going either way and being at times in motion and at times at rest, the rest being either always in motion or always at rest) to be produced by division.[414] Eudemus[415] at any rate, also pretty much paraphrasing the [words] of Aristotle, assumes these sections when he concisely states: 'either all of them go either way or some do and the rest are either in motion or at rest'. But it is clear that if the above threefold division is proved true, the present two-way division is also refuted.[416]

Chapter 4

Translated by István Bodnár

254b7-33 From among the movers and the moved entities [some move and some are moved incidentally, others *per se*. Entities that move or are moved incidentally are e.g. those which belong to the movers or to the moved entities, or which move or are moved according to a part. Entities that move *per se* are those which move or are moved not because they belong to the movers or to the moved entities, and not because a part of them moves or is moved. Of the things that are moved *per se*, some are moved by themselves, some by some other thing, and some naturally, some by constraint and against nature. For that which is moved by itself is moved naturally, as e.g. every animal. (For the animal is moved by itself, and of all the things that have the principle of their

> motion in themselves, these we say that they are moved by nature. Because of this the animal as a whole moves itself by nature, its body, however, can be both moved by nature and against nature. For this depends upon with what motion it happens to move, and what element it consists of. And from among the entities which are moved by something else some are moved by nature, some against nature, e.g. the ones consisting of earth upwards, and fire downwards, furthermore the parts of the animals are often moved against nature – that is, against the position and the mode of movement. And the fact that they are moved by something is most evident in the case of those which are moved against nature, because it is clear that they are moved by something else. Next to those [that are moved] against nature, among those [that are moved] according to nature [come] the ones [which are moved] by themselves, e.g. animals. For it is not unclear, whether they are moved by something, but how the mover and that which is moved have to be distinguished in them. For as in ships and in entities which are not constituted by nature, similarly also in animals apparently the mover and that which is moved are distinguished,] and in this way the whole moves itself.[417]

In order to prove that some entities are always unmoved – which he now formulates as being always at rest, though rest is not said in a strict sense about the entities which are always unmoved[418] – whereas others 1207,1
are always moved, yet others are sometimes at rest and sometimes moved, he divides the moved entities and the movers into those which are moved and which move *per se*, and those which [are moved and which move] incidentally. It has been proved earlier that some entities do move, whereas some others are moved incidentally. Here he subsumes under incidental motion also motion according to a part, when we say that the wholes move or are moved if some part of them moves or is moved, even though in the division he gave of movers and moved entities he contrasted these entities which move or are moved according to some part both with those [which move or are moved] *per se*, and with those [which move or are moved] incidentally.[419] For the ones which move or are moved according to a part, i.e. according to something else, are the opposites of those which [move or are moved] primarily, whereas the ones which [move or are moved] incidentally are the opposites of those which [move or are moved] *per se*.

Here, however, for everything that does not move (or is not moved) in the strict sense, primarily and *per se* he uses the term [moves (or is moved)] 'incidentally'. Having made this first division, he passed over those which [move or are moved] incidentally, as ones which do not contribute for him anything to the investigation of the present topics, and also because those which move or are moved in this way [i.e. incidentally] are themselves referred back to the ones which [move or are moved] *per se*. For those things which are present in the things which move, or are moved *per se* – either as their parts, or as their affections, or as their dispositions, or as their organs, or in some other way – are said to move, or to be moved incidentally.

Having added the entities which move, and are moved *per se*, he made clear what they are by denying the characterisation of being incidental. He divides them in turn into the ones which are moved by themselves, and the ones which are moved by something else. Taking both of these together, he divides them into the ones which are moved by nature, and the ones which are moved by constraint and contrary to nature, since the ones which are moved by themselves are all moved naturally, but among the ones which are moved by something else some are moved naturally, fire, for example, is moved by the entity which changed it, and made it fire, as we shall learn, whereas others are moved by constraint. Consequently, things moved by constraint, and also those non-ensouled entities which are moved naturally, are moved by something else.

First he proves that the entities which are moved by themselves and from themselves – the animals – are moved by nature. For it is agreed that what are moved according to the principle of motion in themselves, *per se* and not incidentally, are moved by nature, and the entities which move themselves evidently have the principle of motion in themselves.

But since the body of the animal, when it performs the motion which is natural to the animal, often does not perform the motion which it is its nature to perform – e.g. if the body is heavy, but the animal moves upward, by jumps or by climbing – the principle of this motion he assigns by a distinction to be somehow in the animal itself,[420] saying that the animal moves then by nature, but nothing hinders that the body of the animal sometimes is moved against its own nature, and
1208,1 sometimes naturally. The distinction, that at this time the body of the animal moves according to nature, and contrary to nature, results from motion of the animal in so far as it is an animal – this is not always the same motion performed, indeed, these are often contrary motions – and further from the nature of the preponderant element in the body. For it is not the body of the animal *simpliciter*. Hence, when the light [constituent] becomes preponderant in the body, and the animal moves upwards, then the motion upwards is natural also for the body, and the body also contributes to the impulse of the animal. If, however, the animal moves downwards while the light [constituent] is preponderant in the body, the body is dragged down contrary to nature.

Having said about the entities which are moved by themselves – that is to say, about animals – that they are always moved naturally when they are moved according to the principle in them, and about their body that sometimes it is moved naturally, sometimes contrary to nature, in what follows he speaks about entities 'which are moved by something else', that of these 'some are moved naturally, and some contrary to nature'.

First he says what the things are which are moved contrary to nature among the entities that are moved by something else. From these we will also grasp the ones which are moved naturally. The entities which are moved by constraint are moved contrary to nature, to the contrary place to which they are of a nature to be moved, e.g. when the earthy ones move upwards, and fire downwards.[421] But parts of animals, too, he says, often move contrary to nature, both contrary to their position and also contrary to the manner of motion[422] – contrary to their position when someone attempts to walk not on his lower limbs but on his hands, contrary to the manner of motion when someone rolls instead of walking. On these occasions, although the animal as a whole is moved from itself and naturally, the motion is contrary to nature for the bodily parts. Having discovered this manner, in which the entities which are moved by themselves, too, sometimes appear to move contrary to nature for their parts, he adds that in the case of the entities moved contrary to nature their being moved by something else – and not their being moved by themselves – is most evident. For in the case of entities moved by constraint it is evident that the mover is different from the moved entity.

Next to the ones moved contrary to nature, where there conspicuously is a distinction between the mover and the moved entity, those of the ones moved naturally which are moved by themselves, as the animals, also have a mover that is evidently different from the moved entity. For, he says, it is not unclear whether animals are moved by something.[423] For it is evident that they are moved by their soul in their animal motion. But it is a matter of investigation how we should distinguish in them the mover and the moved entity[424] – whether as separate in every way, both in [their] nature and place, or in some other manner. For apparently the mover and the moved entity are related to each other as in the case of the non-natural ones, which are moved according to craft and have their mover in themselves, as in the case of ships and chariots. For these, too, have the cause of their motion in themselves – the helmsman and the charioteer – although these are distinct entities and are dependent on their own nature. In the case of animals, too, the mover and the moved entity appear to be distinct according to nature, even if not distinct in place, and so the animal as a whole appears to move itself, and the fact that these are also [caused] 1209,1
by something does not entail that these [movements] are not [caused] by [the animals] themselves.

254b32–255a20 The greatest puzzle is with the remaining part [of the division mentioned last. For we have posited that from among the things moved by something else some are moved contrary to nature, for others it remains that they should be opposed to these, that they are moved by nature. These are the

> ones which would present the puzzle, what they are moved by, i.e. the light and the heavy ones. For these are moved to the opposite places by constraint, into their proper places, however – the light ones up, the heavy ones down – by nature. But by what they are moved, is not clear yet, as when they are moved contrary to nature. For it is impossible to say that they are moved by themselves. For this is characteristic of life, and is a property of ensouled beings, and they would be able to stop themselves. (I mean that if it were the cause of walking, then it could also be the cause of not walking.) Hence, if it is up to fire to be moved up, it is evident that it is up to it also to be moved down. It is not reasonable that they shall be moved with a single motion by themselves if they indeed move themselves.
>
> Moreover, how can something continuous and conjoined move itself. For in so far as it is continuous, not by touch, in so far it cannot be affected. But in so far it is separated, in so far one part is of a nature to be active, and the other part to be acted upon. Therefore none of these moves itself (for they are conjoined), nor does any continuous thing [do so], but in every one of them [i.e. the entities moving themselves] the mover has to be distinct from the moved entity, as we see in the case of entities without soul, when some ensouled entity moves them.
>
> But they also happen to be moved by something, and this would become evident] for those who make a division of the causes.

He has already asserted that the mover and the moved entity are evident both in the case of those which are moved by something else, and those which are moved by themselves, viz. in the case of animals, and even earlier[425] he asserted furthermore that all the entities that are moved by themselves are moved by nature, whereas among the ones which are moved by something else some are moved by nature, others by constraint and contrary to nature. Further, about the ones moved by something else by constraint he has said what these are, and that they evidently have a mover, and a moved entity. He then says, the remaining part of this division is problematic: he means those from among the ones moved by something else which move by nature and naturally. For, he says, only these present the puzzle 'what are they moved by',[426] since it was the case that both the ones which are moved by themselves, and from among the entities that are moved by something else, the ones which are moved by constraint, have the mover and the moved entity in an evident manner. Hence it is quite reasonable that these should present a puzzle since the cause of their natural motion is not clear, 'as the light and the heavy bodies'. 'For when they are' moved 'to the opposite places by constraint,' it is evident that they have a mover, but when [they are moved] 'to their proper places – the light ones up, the heavy ones down, [they are moved] by nature'.[427] But by what they are moved is now not evident, as it is evident 'when they are moved contrary to nature'.

Their case too is problematic, because it seems absurd that they should be moved by something else when they perform their natural motion, since they are said to possess the principle of their motion in themselves (for animals were said to be moved by themselves on this very same account), while to assert that they, [heavy and light bodies], are moved by themselves is to say something impossible. Having proved that this is impossible, it only remains that they are moved by something else, and he investigates what moves them in their natural motion, which is different from them.

He proves by several arguments that it is impossible that these are moved by themselves. The first is the one that it is the sign of life that something moves itself. For we all say that for something to move itself is the same as to live, as Plato says in the tenth book of the *Laws*, '– Are you asking me whether we say that something lives when it moves itself. – Indeed.'[428] And, Aristotle says, it is the property of ensouled beings that they are moved from themselves. This is also said by Plato in the *Phaedrus*: 'that for which motion is from outside – he says – is without soul, that for which it is from within, is ensouled'.[429]

If, then, these bodies [i.e. the elements] are without a soul, and the ones which move themselves are ensouled, these would not be moved from themselves – the inference is in the second figure.[430]

He adds a second argument to this, that if it could move itself, it could also stop itself. For that which is in command of moving itself, does also possess the power to stop itself. For if it has [the power] of 1210,1 walking, then also of not walking, as we see in animals which move from themselves. For just as they walk when they want to, on other occasions they will also stop, whenever they want to.

Not only are the same things in command of stopping, but also of moving with contrary motions – from right to left, and in reverse, and similarly for the other dimensions of place. Aristotle did not add this outright, instead he indicated this in the case of fire, saying 'if it is up to the fire to move upwards, it is clear that it will also be [up to it to move] downwards'.[431] And he indicated this also by what follows, viz. 'it is not reasonable that they shall be moved by themselves only with a single motion if they move themselves',[432] because what are moved from themselves are able to be moved not only with a single and simple motion, but with several, and even contrary motions. Alexander says that he [Aristotle] says here 'and it is not reasonable' instead of 'for it is not reasonable', this clause introducing the cause of what was said earlier.[433]

After this he introduces a third argument proving that the elemental bodies are not moved by themselves with their natural motion. The syllogism goes as follows: the elements are conjoined with themselves, and are continuous in the strict sense, not in a way in which bodies which are in contact, side by side are also said to be continuous in an improper sense. What are continuous in the strict sense are not affected by themselves (this he suggested from the opposite: for it is in virtue of the active and the passive component being separated that one is of a nature to be active, and the other to be affected). What are not affected by themselves would not be moved by themselves. For being moved is a kind of being affected. And the conclusion – that the elements are not moved by themselves – is clear.[434] He adds, further, that 'no other continuous thing either'[435] – reasonably so, since the elements, too, are not affected by themselves due to the fact that they are continuous. For the active and the passive components cannot be continuous, but it is necessary that they touch each other, as he proved in the previous book. For the active component is separated in its substance from the passive one, since the one is in potentiality, whereas the other in actuality. For what is potentially something, is some other thing in actuality, for fire exerts its activity into air, which is fire in potentiality, and air in actuality. 'How do animals move themselves', Alexander asks, 'if they are continuous?' And he gives an obvious solution, that in animals the mover is not continuous with the moved part. For the soul is not a body, and neither is it a quantity at all. Because of this, they do not touch each other. And Aristotle, too, said a little while ago that as in vessels, so in animals, the mover and the moved part are distinct from each other by nature. That the mover has to be distinct as against the moved, can be seen in an evident manner about things without a soul when they are moved by ensouled beings. The ensouled beings, too, he says, are always moved by something,[436] for everything that is moved is moved by something, and the mover is different, distinct from the moved part, and not continuous with it, as it is not even a quantity.

I think Alexander is right to add that Aristotle may have said 'but these, too, happen to be moved always by something' not about ensouled beings, but about things which are moved naturally. For these, too, will be proved to be moved with their natural motion by something, even in the cases when their mover, which moves them with this motion is not apparent in a way in which one can see the mover which moves [them] with a constrained motion. As we see in their case that the mover and the moved entity are distinct, so it will be proved that the natural bodies are moved by their natural motion by some distinct causes. And this, he says will be apparent for those who make a division of the moving causes.[437] For he will prove that some of the movers move by contact, being contiguous with what they move, whereas others move by making actual what previously is in potentiality – for example water generated in actuality from air, being no longer able to remain in the place of air, moves to the proper place of water, being in some sense moved by what made it water in the first place.

255a20-30 What has been said can be grasped also in the case of the movers. [For some of these are productive of motion contrary to nature, e.g. the lever is not productive of the motion for the load, whereas others are [productive of motion] by nature, e.g. the hot in actuality is productive of motion for the hot in potentiality. And similarly for the other such things. And that which is potentially of such quality, or [potentially] of such quantity, or [potentially] somewhere is receptive of motion by nature, when it has such a principle in itself, and this not in an incidental manner – for the same thing may be of such a quality and of such a quantity, but one of these would be incidental to the other, and would not be present *per se*.

So fire and earth are moved by something – by constraint when [they are moved] contrary to nature, by nature when [they are moved] into their actualities,] they being in potentiality.

As he intends to prove that from among the bodies without soul the ones that move with a natural motion are also moved by some external entity, and not by themselves, he establishes this by dividing the movers, just as the moved entities into those that move naturally, and those that move contrary to nature. If this is true also of these [movers], it is clear that those that are movers of the movement contrary to nature move entities which are moved by constraint and contrary to nature, and those that are movers of the natural movement [move] the [entities which are moved] naturally. He introduces the lever as one among those that move contrary to nature. For none of the entities that are moved naturally are moved by the lever, but it is by constraint, the way such movement is,[438] for things that have weight are moved up by the lever, with the motion contrary to their nature, and the lever moves [the load] contrary to nature because it does not move according to its internal principle – that way it would move it down – but it receives the cause of motion from outside, from some different entity. Such are movers that move contrary to nature.[439] What is thus-and-so in actuality is said to be the mover naturally of what is such in potentiality.[440] For the things which are moved in this way the mover which moves in this way is the cause of natural movement, and it moves according to its internal cause and principle. For what is warm in actuality, when it warms what is at that time warm in potentiality, moves this way, because it has the cause of motion in itself.

He sets alongside this that which is moved by nature, and so makes the discussion clearer. For just as what is in actuality is the mover by nature of what is in potentiality, what is in potentiality can be moved by nature by what is in actuality. And just as what is the mover in potentiality possesses in itself the principle and capacity of movement, so it can also be moved by nature, when it has the principle of movement in itself, now not exerting activity, as the mover by nature has it, but according to suitability and capacity – and so it is potentially of such quality, or of such quantity, or potentially up or down. He mentioned

1212,1 these categories because he thinks that motion comes to be only according to these. Saying 'when it has such a principle in itself' he adds 'and this not in an incidental manner'.[441] For even if what is potentially of this quality is also potentially of this quantity, nevertheless it changes into this quality *per se*, inasmuch as it is potentially of this quality. Since it is also potentially of this quantity, if someone were to say that what is potentially of this quantity changes into what is of this quality, such a change would be one in an incidental manner, because it happened to be also potentially of this quantity. What is potentially of this quality changes *per se* into [what is of this] quality, and what is potentially of this quantity [changes *per se*] into [what is of this] quantity, whereas the crosswise change happens incidentally.

He explained what the entities are that are moved naturally through the foregoing arguments, formulating his account in terms of the three categories. But he does not add then what the entities are that are moved contrary to nature in terms of the three categories. In the case of locomotion, as each body has both a natural motion, and one contrary to nature – the one which is into the contrary [place] to the natural one, the entities which are moved according to place contrary to nature will be evident. For they do not possess the principle of such motion in themselves. In the case of qualitative change and increase, the change contrary to nature in these respects is no longer evident. Or perhaps he has also mentioned the motion contrary to nature belonging to them in the previous examples, in the case of increase of corn mentioning that in the so-called gardens of Adonis[442] the corn grows contrary to nature through some such suitability, and for this reason they do not reach maturity; and those people who reach puberty or old age quickly because of their licentiousness will both grow and diminish contrary to nature. For they will also change their temperament contrary to nature, as one becomes warmer, and another one colder more quickly than it is right. But his inquiry does not concern these now, but rather the simple bodies – by what are they moved with their natural movements?

Having said what the entities are like that move by nature, and [what the ones are like] that are moved [by nature], he introduces finally what was the purpose to which he embarked upon this argument, and says that earth, and fire, and in general the simple bodies, whenever they are moved, are moved certainly by something else – by constraint when they are moved away, contrary to nature, from their proper places to the contrary ones, by nature when they are moved to their proper places, to which they are of a nature to be moved naturally. He says: 'whenever they, being in potentiality are changed previously into their own actuality',[443] that is to say when they become in actuality what[444] they were previously potentially, and when they acquire their proper actuality. For fire, when it becomes fire in actuality, having been [fire] previously in potentiality, is then active with the actuality of fire and moves with the natural motion of fire. He does not mean by 'into

their actuality'[445] that they are changed 'into [what they are] in actuality according to substance' – for this is natural generation, and not motion – but when having been generated by something it acquires its proper actualities, according to which it is moved thereafter naturally by that entity, being carried to its proper place. For he says the 1213,1
actuality, and this movement to its proper place is provided to it by that entity which gives it its substance.[446] And this is what is meant by [the claim that it is] moved by that entity.

Alexander says that he calls the natural places of the bodies their 'actualities', i.e. [their] perfections. 'For he thinks', Alexander says, 'that each of these bodies is heavy, or light, in actuality when it is in its proper place. For it is not said to be potentially fire, or water, or earth, or air, before it is in its proper place. Because if they were potentially one of these [before arriving at its proper place], they would not be moved, but rather generated.' But these are in actuality, even when they are in the place belonging to some other body, and they receive the perfection according to place, when they are in actuality according to what each of them is. The actuality of the light entity is to be moved up, for it will be moved towards that in which it is in potentiality – just as the entity which is warmed. Hence both movement and change towards perfection will be natural, and this perfection is nothing but to be in its proper place. Accordingly, if I am not mistaken, according to him [Alexander], too, it is not the proper place which is the proper actuality – as he started to assert – but the movement towards the proper place. Therefore, those things move by nature which acquire their proper form, according to which they previously were in potentiality, and acquire together with the form in actuality the actuality which is according to this form.

255a30-b24 As 'potentially' is said in many ways [this is the cause why it is not evident by what such things are moved – e.g. fire up, and earth down. He who learns knowledge, and he who already has it, but is not in actuality, are potentially in different ways. When that which is productive, and that which can be affected are together, that which is potential will always become actual, as that which learns from being potentially becomes different in potentiality – for he who possesses knowledge but does not contemplate, is a knower potentially in some way, but not in a way in which he used to be before he learned. And whenever he is so, if nothing hinders, he is in actuality and contemplates, or else he will be in the contradictory state and is ignorant.

And these things are similar also with the natural entities. For the cold is warm potentially, and when it has changed, and is already fire, it will burn, if nothing hinders and obstructs. And similarly with the heavy and light. For the light comes to be from the heavy, e.g. from water air – for this is first potentially – and it is already light, and will be in actuality immediately if nothing hinders. The actuality of light is to be somewhere – to be up – and it is hindered when it is in the opposite place. And similarly with [a thing] of some quality and of some quantity.

Nevertheless, this is being inquired into, why are the light and the heavy things moved to their own place. The cause is that they are of a nature [to be] somewhere,[447] and this is what it is to be light and heavy, [light] being determined by up, and heavy by down.

Light and heavy are in potentiality in many ways, as it has been said. For when it is water, it is light potentially in some way, and when [it is] air, in a way it still is potentially – for it is possible for it to be hindered, and so not to be up. But when the hindrance is removed, it is in actuality and always gets further up. And [what is] of some quality changes into being in actuality, for that which has knowledge contemplates immediately if nothing hinders it.] And [what is of] some quantity will expand, unless something hinders it.

Having said that those entities move naturally, which move into their actualities as they are potentially, either because they are potentially in their proper places, or that they still have something potentially, because motion is the actuality of the potential in so far as it is potential, he adds that this 'potentiality' is 'said in many ways' in the moved entities, 'and this is the cause why' the issue being investigated at present is not clear, i.e. by what are such things moved, e.g. fire up and earth down.[448] For that which is moved by something has to be subject to motion itself, and [has to be] potentially, on the other hand that by which it is moved has to be something that causes motion, and [has to be] in actuality; in fire and earth this potentiality is not clear, because 'potentially' is said in many ways, that which causes its motion, and is in actuality, will also be unclear.

Now, 'potential' is divided into (1) that which is according to suitability merely, although it has not yet become in any way such as it is said to be (as we say that a little child knows grammar, or music potentially, because he is able to become such), and also into (2) that which has become such, and possesses the complete disposition, but is not active according to it, as we say that Aristoxenus knows music potentially, when he does not play the lyre, but sleeps or is active in some other thing. And this potentiality is in between the things that are completely potentially, and only potentially, and the ones that are completely and only in actuality, because it partakes of both, and is related to each of them as the other [counterpart].[449]

Having proved that the physical bodies possess these potentialities in natural motions, he proceeds to discover what are productive of movement for both of these potentialities.

First he says how and by what the change into actuality occurs in the case of both potentialities. For whenever – he says – that which is productive of action, and that which can be affected are together, the first potentiality, the one which is not completed, changes by the actuality as a next step into the actuality of disposition, which is itself still in some sense potentially, according to the other potentiality of being not yet actual. Of the actuality of this potentiality – so that it is completely actually – the disposition is also the cause, but the person removing the obstacle is also responsible, as will be said, and that which did produce the disposition in it is first and foremost the cause also of this [i.e. of the actuality of this potentiality]. 1214,1

In the phrase stating, 'always when that which is productive, and that which can be affected are together, that which is potential sometimes comes into being actually' Alexander says that the word 'sometimes' is not added in some manuscripts. Even if it is added, it expresses firm truth, first because even if that which is productive is together with that which can be affected, if that which is productive is not completely actual – so that it possesses the disposition and is actual without hindrance – that which is potential will not come into being actually. He [Alexander] himself, as he interprets the word 'sometimes' says 'whenever it is so, if nothing hinders it, it is actual and contemplates'.

Furthermore, in the case of bodies and bodily affections the fact that that which is productive and that which can be affected are adjacent, and become together, is sufficient for that which is potential for it to come to be actual. In the case of things to be learned and taught, however, the fact that that which teaches and that which learns come together is not sufficient for the [teacher] to teach and for the [learner] to learn, if there is no deliberation and time. So the addition of the word 'sometimes' is useful.

It is, however, obvious that that which changes from the previous potentiality into the disposition changes certainly by something else which is actually. This is the result of generation, and not of motion in the strict sense.

When, he [Aristotle] says, it has acquired the disposition, if nothing hinders it, it is actual and contemplates,[450] joining the actuality already to the disposition. For being hindered by nothing were it not to be actual, it would be 'in a contradictory [state], and not in ignorance without further qualification', as Alexander writes. For if he has changed into knowledge, and has the disposition of knowing from the teacher, and nothing hinders him, nevertheless he is not in actuality in respect of knowing, such a person 'is not in ignorance without further qualification', but 'in a contradictory [state]'. For he would be knowing in so far as he has knowledge, and not knowing in so far as – even though nothing hinders – he is not actual in respect of knowledge.

It is clear that it is not impossible to be 'in ignorance without further qualification', whereas to be 'in a contradictory [state]' is impossible. If, however, the reading was – as it is in most of the manuscripts – that 'he will also be in a contradictory [state] and in ignorance', what is said will more or less be the same. [This reading] says that he who has the disposition, and is knowing according to the disposition, but is not in actuality although nothing hinders him, he will be in a contradictory [state], because he is both in ignorance and in knowing.

1215,1 Alexander was aware of this reading, too. If it is impossible that that which has the disposition from something is not in actuality according to the disposition, if nothing hinders it, but actuality and the motion, with which it moves according to the disposition, are conjoined to the disposition, it is reasonable to say that it is moved by that which did produce the disposition in it.

Having thus demonstrated the change from the two potentialities into the actuality, and indicated that the actuality is conjoined with the second potentiality, if nothing hinders it, he passes over to the natural bodies, and proves that in their case also each of them is in general moved with the natural motion in it by what changes it from the first potentiality into the second potentiality, because the actuality is readily present, if nothing hinders the disposition. For when warm comes to be from the cold, and fire from the water, it is thereafter active, in so far as it is fire – and it burns, and is moved up, if nothing hinders it. And so it has this activity from what has produced it to be fire in actuality. For it does not receive in addition [the capacity of] burning and not burning[451] after it has become fire, but its actuality is linked to its change into fire, which it has from what has produced it.

Having proven that this is so in general in the case of the physical bodies, how the change comes to be from the first potentiality into the second, the one related to disposition, and how actuality and motion are conjoined with this, if nothing hinders, next he divides the account into the different motions. He first presents the account concisely, afterwards in an expanded fashion. For, he says, from the actually heavy but potentially light, the light comes to be according to the first potentiality. For instance, from water – which is actually heavy, and potentially light – air comes to be, which is actually light, and 'it will be in activity immediately, if nothing hinders it'. Alexander understands the clause 'for this is first potentially, and already light' to refer to the incomplete potentiality, which is heavy according to actuality. 'For', he says, 'that which is such, and is in such a state that it is still heavy in actuality, is light potentially according to what is called first potentiality. Once it has changed, and has become light, it immediately is active with the activity of the light'.

Nevertheless, I do not think that Aristotle says the phrase 'and already light' about that which is heavy according to actuality, and is light potentially, with reference to the incomplete potentiality. For the phrase 'already light' does not accord with this, but to what has already become light according to actuality. Neither does what is added – that 'it will be in activity immediately, if nothing hinders' – accord with incomplete potentiality, but to the potentiality according to disposition. If then this whole [phrase] – that 'for this is first potentially and is already light and will be in activity immediately, lest something should hinder [it], but it is already light and will be in activity immediately' – is said about the same thing, it is appropriate to the more complete potentiality, the one which is according to disposition, one has to understand the clause 'for this is first potentially' to refer to the same thing. 'First' does not indicate the order of generation, in which the less complete is the first, but [indicates] completeness [instead], for the dispositional potentiality is more complete 1216,1 than the one according to suitability – 'the actuality', he says, 'of the light' is to be in the upper place: for this is what the phrase 'somewhere and up' means, and also the fact that it rises to the surface of the others, for this, too, is an actuality. And clearly, the actuality of the heavy is to be down and to settle under the others.

Strictly speaking the actualities of the light and the heavy are to be up and down, because they acquire then their actualities in the strict sense and their completeness according to form. For the movement up and down belongs to them while they are still incomplete, and have not yet acquired their completeness according to form.

Having spoken about motion according to place 'the case is similar', he says, 'with quantity and quality'. For that which is increased, when it is made as great in actuality, as it was in potentiality, by that which increases it, then it immediately is active with the actuality according to what it is so great. It expands – indeed, it rather has expanded and distended in actuality to be as great, as was its size in potentiality, and this is its complete actuality, having been expanded and distended to be this great. [The process of] expanding and distending belongs to it when it is still incomplete, and consequently [this process] is an incomplete actuality.

'And similarly quality, too' changes from the incomplete potentiality to the complete, i.e. to the disposition, and the complete actuality follows immediately. For 'that which knows contemplates, if nothing hinders it', and this is the complete actuality of the knowledge in actuality. For learning is still an incomplete actuality, and it is incomplete.[452]

Having said that the moving cause of those things which move naturally is what makes them actually from the incomplete potentiality, he reminds us that the object of inquiry was through what and by what cause do the natural bodies move to their proper places, which has now been discovered not only for the case of things moved according to place, but also for the ones [moved] according to quantity and quality. For the cause is that which has made them to be such that this kind of motion will follow of necessity to them. 'The phrase "and this is being inquired into", Alexander says, can be meant not only of what is the puzzle now, but also as of something inquired into in a general manner'. For in actual fact it is being inquired what is the cause of the locomotion

to the proper places of the bodies, and the cause has been discovered, that they are 'of a nature' to be 'somewhere'. For this is 'what it is to be heavy and light', that one of them is up, the other is down, and if they were immediately so as they are when they are complete, having their completeness from themselves, they would be the causes also of their own actuality. But since before they are in actuality so, they are first such in potentiality, the potentiality [in this case] is also twofold, as has been said.[453] For when it is water, being heavy in actuality, it is somehow in potentiality.[454] For it is air according to the incomplete potentiality, and because of this, is light [according to the incomplete potentiality], and when it becomes air, even then it is in some way still [light] in potentiality, because it is sometimes hindered and is not yet actual according to the actuality of its form, according to which the light
1217,1 body is up. When, however, the obstacle is removed, it is active always advancing from its incomplete actuality to the more complete one, and becoming always further up, until it gets uppermost, and acquires its most complete actuality, that it is uppermost.

He says that quantity and quality also change into actuality in a similar manner, if nothing hinders. For there are some things which have been increased and have acquired the form, nevertheless they do not yet occupy so much place as they are able to occupy, because they are hindered in this, as the air generated from water in a small room. For it would expand further, if it were not compacted and compressed by the room.

> **255b24-31** He who moves the obstruction and hindrance [moves in one way, and it does not move in another way, e.g. he who tears down the column, or removes the pebble from the wine-skin in the water. For he moves in an incidental manner, just as the rebounding ball is not moved by the wall, but by him who has thrown it. It is evident that none of these moves itself. But they have a principle of motion, but not of moving,] nor of producing [motion], but of being affected.

Up to this point, as it seems to me, he has said that the cause of natural motion and actuality is the completeness of the form: once something acquires it, it will be active *per se*, if nothing hinders. For the assertion that it is moved in this way due to having such a nature pertains to someone who specifies as the cause the nature and the form.

Advancing somewhat further, he says that what has generated and made it to be such as it is, is also the cause of such motion, and the interpreters supposed that he has specified this also in what was said earlier. On several occasions, however, saying that as it acquires the disposition it is active immediately, he adds 'if nothing hinders'. Now he says that 'he who removes the obstruction and hindrance[455] moves in some way, and in some way does not move'. He does not move *per se*, for he is principally active in something else, and he moves incidentally. 'For he who pulls down the column' moves it *per se*, incidentally, however, it also moves the roof supported by the column, by which it was hindered to be brought[456] down according to its nature. In a similar manner, he who removed the stone which is on the inflated wine-skin, due to which it was submerged, and did not rise to the surface of the water, principally he moved the stone, but incidentally also the wine-skin, which – once the stone which held it down was removed – is carried up.

He proved even more the dissimilarity between the motion which is produced by entities that move *per se*, and the ones which remove the hindrance of motion, with the ball which bounces off the wall. For just as this is moved principally by the person who threw the ball in the first place, it is also moved incidentally by the wall, from which it bounces off, similarly the physical bodies are moved principally by what changed[457] them and made them such, in an incidental way, however, [they are moved] also by what removes the obstacles to their motion.

As it follows from what has been said that none of these [elements] are moved by themselves, but by something else, an objection presents itself, asking how these physical [bodies] are said to have in themselves their nature as the principle of motion, if they are not moved by themselves, from within, but by something else, from outside. And he resolves this objection by saying that these are said to have 'the 1218,1
principle of motion' not as [the principle] 'of causing motion, nor of producing [motion], but of' being moved and 'being affected'. For not only that which moves from itself is said to possess a principle of motion, but also that which is of a nature to be moved, is said to possess a principle of motion. For the term 'motion' is common both to the mover and to that which is moved. If motion is in that which is moved – as it has been proved earlier[458] – motion is more proper to that which is moved. Physical bodies are said to possess the principle of being moved by the fact that they possess the capacity and aptitude to change into that, changing[459] into which they will be moved by something with the motion the principle of which they are said to possess. For to be of a nature to be moved with some motion is the same as possessing the principle of such motion, as e.g. one who is gifted for philosophy [is said] to possess the principle of philosophy, not [the principle] of making someone else a philosopher, but [the principle] of himself becoming a philosopher.

If, then, the nature of these bodies is not a principle of causing motion and producing [motion], but of being moved and of being affected, when Aristotle says that nature is a productive cause, somewhere saying that 'neither god nor nature does anything in vain',[460] and somewhere else proving that nature is a productive cause according to reasons and for the sake of something, either we have to think of that nature as some other nature – the demiurgic intellect[461] pervading everything – or, provided he speaks about the same nature, one has to understand what he says in a material and passive way, as we have understood also the principle.

Alexander also inquires about these how the body which moves in a circle will still be natural, if the natural [body] – as Aristotle says – possesses the principle of being affected, whereas the body which moves in a circle cannot be affected. He solves this first by saying that even if the body which moves in a circle is always moved, nevertheless it has potentialities in itself as at different times it is moved from one [part] to another. And in so far as it takes part in potentiality, it can also be affected in some way. For every potentiality is somehow tied to matter. This [body], then, also possesses in itself the principle of motion in respect of being moved, and it is a physical [body] in respect of this. 'Or it is better', he says, 'to understand motion in the divine [body] not as natural motion as in bodies without soul, but as [natural motion] in those which are moved according to soul, which are not moved from outside by something, but possess the moving principle and [moving] cause in themselves.'

'But if this is the case', he says, 'how come it does not possess the principle of coming to rest, and of motion in the contrary direction?' 'Or is it the case', he says, 'that it is not moved with the contrary motion because there is no motion contrary to the circular motion and that it does not rest, because it is not moved to a place in which it would be necessary that what has come to be there stops, and it does not need to come to rest. Those entities which are moved by themselves and are of a nature to stop, these also possess the capacity to stop by themselves.'

This is what Alexander says. What he means when he says that in the case of the heavens natural motion is not to be taken as in the case of bodies without soul, but as [the natural motion] in the case of those which are moved according to soul, we shall know more clearly when
1219,1 we have commented on the doctrine which he has about the soul of the heavens. For he thinks that soul and nature are the same in the heavens. He indeed writes in the commentary on the second book of *On the Heavens* the following: 'We intend to prove that the nature and the soul of the divine body are not different, but like the weight of earth and the lightness of fire' and shortly afterwards 'For what other nature would it have besides this [the soul]? For soul is a more complete nature, and it is reasonable that the nature belonging to the more complete body is itself more complete.'

This is then what he said also here – that the natural motion of the heavens is like [the motion] of entities moved according to soul. It is clear that this he says in accordance with his doctrine about the soul, as he thinks that the soul is an inseparable actuality of bodies, just as nature is also such [an inseparable actuality of bodies].

255b31-256a3 If everything that is moved [is either moved by nature, or contrary to nature and by constraint, and everything [that is moved] by constraint and contrary to nature [is moved] by something, and by some different thing, and again, from among

> the things moved by nature the ones which are moved by themselves are moved by something, and the ones that are not moved by themselves – e.g. the light things, and the heavy ones are moved either by what has produced them and made them light or heavy, or by what disposes of the hindrance and obstacle,] everything that is moved would be moved by something.

Having proved immediately preceding that the things said to be moved by nature are not moved by themselves, but by something else,[462] and regarding it as accepted that the things moved by constraint and contrary to nature are moved by something (for this is evident), and taking it that also in the case of animals the fact that they are moved by something is evident and accepted, and that the only object of inquiry in their case is how the mover and the entity being moved have to be distinguished, taking all these points as having been accepted earlier, he now concludes from what has been posited that everything that is moved is moved by something. For if some of the things moved are moved by nature, and some are moved contrary to nature and by constraint, and the [ones that are moved] by constraint and contrary to nature are all moved by some other external mover, and from among the [ones that are moved] by nature some are moved by themselves – as the animals – and some are not moved by themselves – as the light and the heavy things, and in general the simple bodies,[463] about which it has been proved that *per se* [they are moved] 'by what has generated them and made them light or heavy',[464] and in general such that once they are generated they move up or down, whereas [they are moved] incidentally by that which removes the obstacles and hindrances, it is clear that everything that is moved is moved by something.

For there is nothing that is moved beside these – the ones [moved] contrary to nature, and the ones [moved] naturally, and the twofold distinction concerns these. The syllogism is the following: Everything that is moved is moved either naturally or contrary to nature, everything that is moved contrary to nature and according to nature is moved by something. Therefore everything that is moved is moved by something.[465] One can conclude also in a hypothetical syllogism, as Aristotle himself seems to conclude. For if everything that is moved is moved either by nature or contrary to nature and both the [things that are moved] by nature and the [things that are moved] contrary to nature are moved by something, everything that is moved is moved by something. The first – for it has been proved – therefore the second.[466]

He has proved here in a more physical manner than in the preceding book[467] that everything that is moved is moved by something. Eudemus
proves that that which is moved is moved by something also from the 1220,1
fact that relatives subsist together. 'For as it is not reasonable – he says – that that which is cut is not cut by something, and that which is dragged is not dragged by something, and in general that which is made

is not made by something, so [it is] also [not reasonable] that that which is moved is not moved by something'

I think that one might pose the puzzle, how does [Aristotle] assert that the entities which are moved naturally with locomotion are moved *per se* by their mover, if that which generates and produces [them], produces their substance and form, and is the cause of their substantial change, whereas substantial change and locomotion are different? And in general, if that which generated and made fire stops at some time, and is not present and does not touch that which is moved, how is fire said to be moved by it, so as actually to conclude that everything that is moved is moved by something else apart from them?

Let us investigate this further, but let us say meanwhile what now comes to mind, that in some way that which produces the form of that which is of a nature to move with this motion, is also something that causes motion. For it produces the motion and the actuality of the form, together with the form, as he who produces the vase also produces that it receives wine. But he who produces the motion together with the form strictly speaking does not move the form with this motion, as he who moves the stone with his hand – since the producer is no longer present to the entity as it is being moved, but this is apparently a different distinction of that which moves, according to which it produces the form to be of a nature to be moved, also to be moved if nothing hinders.

And he has discovered in a remarkable way also in the case of the entities moved naturally that they are moved by something, since to be moved is to be affected, and that which is affected requires something that is active. For the complete actuality, which proceeds from the complete substance does not require any other creative cause, but motion, being an incomplete actuality, rather an affection, and mixed with much potentiality requires that which produces it.

Chapter 5

Translated by Michael Share (1220,27-1232,32) and Michael Chase (1232,33-1250,31)

256a4-21 And this in two ways: either (i) not because of[468] the mover itself [but because of another thing which moves the mover, or (ii) because of the mover itself. And this latter is either (a) first after[469] the last thing [in the series] or (b) [acts] through a number[470] of intermediaries, as when the stick moves the stone and is [itself] moved by the hand, which is moved by the man, but he no longer [moves it] by being moved by something else. We say, then, that both cause motion – the first and last, [that is,] of the movers – but more so the first. For that moves the last, but not the last the first, and without the first the last will not cause motion, but the first will without it. For instance, the stick will not cause

> motion if the man does not move it. If, then, everything that is in motion must be moved by something, and either (1) by something that is moved by something else or (2) not, and if [by something that is moved] by something else, there must be some mover which is not first [moved] by something else, and if the first mover[471] is such, there is no need of another (for it is impossible for a series of movers that are themselves moved by something else to go on to infinity, for nothing is first in an infinite series)[472] – if, then, everything that is in motion is moved by something, and the first mover is moved, but not by something else,] it must be moved by itself.

Having shown[473] that everything that is in motion[474] is moved by something, he would next, following due sequence, have asked[475] whether the thing that is in motion is moved by something that is in motion [itself] or by something not in motion (*akinêtos*); but because, as Eudemus says,[476] it seems to practically everybody that a thing that causes motion is also in motion itself, and given that, of things that cause motion and are at the same time in motion, some are self-moved, others moved from outside by something else (there are two things that are originative of motion: the unmoved and the self-moved), he first introduces the self-moved and shows that among things that both cause motion and are in motion [themselves] there is something primary which causes motion without being moved from outside by something else.

He divides that which causes motion *per se* once more,[477] using a 1221,1 different division, into (a) that which causes motion not because of itself but because of something else that moves it from outside and (b) that which causes motion because of itself. And he says 'because of something else' is as when a thing moves something by being moved by something else, as a lever moves a weight or a stick moves a stone because something else is moving it. On the other hand, a thing that has the cause of motion (*tou kinein*) within itself causes motion 'because of itself', as living creatures cause motion in accordance with their respective drives.

And he further divides that which causes motion because of itself into (i) the first mover after the last [thing in the series], which is only moved – as when a man, causing motion because of himself in that he has the [source of the] motion within (*para*) himself, moves a stone while being first after the last [thing in the series], which is the stone, which is moved but no longer moves anything else, with the result that the stone is moved immediately by the man. Into this, then, he divides that which causes motion because of itself, and (ii) [the case of] moving the last [thing in the series] through a number of intermediaries, like a man who, while causing motion because of himself, moves a stone, which is *ex hypothesi* the last thing moved, through a number of intermediaries, as for instance when a man moves his hand, his hand a stick (let us agree to count the hand too as an intermediary), the stick a stone, and the man does not cause motion by being moved by something else and the stone is moved without itself moving any further thing.

And although, he says, both [types of things] that cause motion in this way cause motion, [i.e.] both the one that is next to the last [thing in the series], like the stick, and the first mover, like the man, the first [mover], i.e. the man, is more, and more strictly speaking, productive of motion, since he also moves the stick but the stick does not move the man, and without the man the stick will not cause motion, but the man will also cause motion even without the stick. For these reasons the man causes motion more than does the stick, since the man does not need anything else moving him to cause motion, while the stick does.

And so everything that is moved by something has need of something that causes motion because of itself, since things that do not cause motion because of themselves but are moved by other things and cause motion in that way certainly have need of something that causes motion because of itself and would not cause motion otherwise, whereas things that cause motion because of themselves do not have need of things that do not cause motion because of themselves. So everything that is moved must be moved by something that causes motion because of itself, and either immediately or through the mediation of one of the things that are moved by something else; and if immediately, i.e. if the thing that moves because of itself comes straight after the thing that is only moved, there is no need for there to be yet another thing serving as an intermediary between the thing that causes motion because of itself and the final moving thing; for something that causes motion because of itself, if it is adjacent (*prosekhês*) to the thing that is moved and first after it, is enough to move it. So if it is necessarily the case that a thing that is moved by something will be either directly or mediately moved by something that causes motion because of itself, i.e. by something self-moved, it is necessarily the case that the first of the things capable of causing motion is the self-moved.

And he further shows that something of this kind, something that causes motion without itself being moved by something from outside but [only] by itself, actually exists.[478] For if everything that is moved and thereby causes motion is going to be moved by something else from outside, and that [in turn] by something else, and this is always the case, it must go on to infinity. But that is impossible. For, he says, it
1222,1 will not be possible to identify (*lambanein*) the first mover, and if there is not a first mover nor could there be a last, the one adjacent to that which is moved but no longer causes motion. So if there is neither a first nor a last mover, there is no motion, since if there is not a mover nor will there be anything that is moved. Therefore one must not go on putting one moved mover before another to infinity. Therefore there is a first mover not moved by another [mover].

Next he assembles his conclusions. It has been shown that everything that moves is moved by something; and [to this] it has been added that a thing that is moved by something is moved by a first mover; to which it has been added that the first mover is moved but not by something else, for it would no longer be 'first' and movers and things moved[479] would have to proceed to infinity and motion would thus be eliminated. And consequently everything that is moved is moved by a first [mover], the self-moved [mover]. For if this is eliminated so are the things that do not cause motion in the same way (*houtô*), but if they are eliminated, it is not also eliminated. And, generally, a thing that is moved by something else is later [in the series] than a thing that causes motion by being moved itself,[480] so the latter [comes] first. Also, the mover is the cause of that which is moved being moved, and that which is caused is later than the cause, so that which first causes motion by being moved [itself] is that which moves itself.

Eudemus adds[481] that among motions here [below] the motions associated with life appear to be causes and that the heavenly [motion] is of course [the cause] of the transformation of the elements. But the latter too seems in some sense to involve life, for the heaven is not moved by something else but by itself.[482]

256a21-b3 It is also possible to reach this same conclusion in the following manner. [Every mover both moves something and moves it by means of something.[483] For the mover either causes motion by means of itself[484] or by means of something else, as a man, for instance, does either by himself (*autos*) or by means of a stick, or as the wind either overturns [things] by itself or a stone it propels does. But it is impossible for that by means of which [a mover] causes motion to cause motion without [there being] something that causes motion by means of itself,[485] but while if [a mover] causes motion by means of itself, there does not have to be something else by means of which it causes motion, should that by means of which it causes motion be something else, there will [always] be something which causes motion not by means of something [else] but by means of itself, or [the series] will go on to infinity.[486] So if something causes motion by being moved, [the series] must stop and not go on to infinity. For if the stick causes motion by being moved by the hand, the hand is moving the stick. And if something else causes motion by means of [the hand], it is also something else that is moving the hand. Whenever, then, one thing after another causes motion by means of something [else], there must be something earlier [in the series] that causes motion by means of itself. So if this thing is moved, but what moves it is not another thing, it must move itself. And so, according to this argument as well, a thing which is moved is either moved immediately by a thing which moves itself], or such a thing is reached at some point [in the series].

[Aristotle] himself says that this argument is the same as the one before it. 'But it differs', says Alexander, 'in its method [of reasoning], because there the starting-point of the proof is [the proposition] that a thing that is moved is moved either by something that moves [it] by means of itself or [by something that is moving it] not by means of itself but by something else, while here his proof proceeds from the mover; for it[487] assumes that just as every mover moves something (it is impossible for something to cause motion but move nothing), so too does it cause motion by means of something'.

But perhaps[488] it is not at this point that the arguments differ – in fact the one already formulated[489] also argues from the mover when it says 'either not because of the mover itself but because of another thing which moves the mover, or because of the mover itself'[490] – but it is by the difference between causing motion 'because of itself' and causing motion 'by means of itself' and between [causing motion] 'because of another thing' and 'by means of something else' that the arguments differ.[491] And these differ in that something that causes motion because of itself causes motion by having the principle of motion in itself, while something that causes motion by means of itself [does so] because that which has the principle of motion in it is in immediate contact with the thing that is moved. For this is the meaning of his statement 'as a man, for instance, does either by himself or by means of a stick, or as the wind either overturns [things] by itself or a stone it propels does'.[492] So, just as he earlier followed that which causes motion because of another thing back[493] to that which causes motion because of itself and showed that if there is something that causes motion because of another thing, something that causes motion because of itself must come before (*pro-huparkhein*) [it], so too does he [here] follow that by means of which the mover causes motion,[494] when it is intermediary between the first mover and the last thing moved,[495] back to that which causes motion by means of itself and show that 'whenever one thing after another causes motion by means of something [else],[496] there must be something earlier [in the series] that causes motion by means of itself', which also has the principle of motion in itself and is adjacent to the thing that is moved [by it] and does not move it through something else.

This, then, is how I think one should distinguish the arguments, even though Eudemus brings them together when he says[497] 'that, then, which causes motion does so either because of itself or because of another thing, or (*kai*)[498] by means of itself or by means of something else'.

Having assumed that everything that causes motion does so by means of something, he [next] divides 'causing motion by means of something' into 'causing motion by means of itself' and '[causing motion] by means of something else'. After all, someone who moves something through himself and not through an intermediary causes motion by means of something, as does the man when he moves the stone by himself, for he moves it by means of his body, which is in contact with it. But even should he move it through a stick, he is causing motion by means of something then as well, for the stick is also 'something'.[499]

Next he adds to this division [the proviso] that it is not possible for that through which [something] causes motion, if it is some other thing,[500] to cause motion without [the involvement of] that which causes motion by means of itself. For unless the man, who causes motion by means of himself, imparted motion, the stick would not ever move the stone. Indeed, it would not even begin to move. For that which causes motion by means of itself does not necessarily also have need of anything [further] with which to cause motion. For it is competent to do this even on its own, while the intermediary by means of which something is moved in every case has need of something that causes motion by means of itself so that, having been moved by something else, it may cause motion. For if there were not something of this kind, but any given thing was always moved by something else which was itself also moved from outside, and this was always the case, there would once more be a progression to infinity of things that both cause motion and are moved. But that would be absurd, both *tout court*, since if there were not a starting-point for the motion, there would not even be any motion, and obviously also on account of what was shown[501] in the preceding book. For there will be infinite motion in a finite period, since the magnitude of what is being moved will become infinite. For if the movers and things moved are infinite [in number], a single magnitude must result from them, if not a continuous one, one [formed] by their being in contact, since all things that cause motion in place by virtue of being moved themselves do so through contact. Therefore [the chain of unmoved movers] must come to a halt.

And so if everything that causes motion does so by means of something, and either by means of itself or by means of another thing, and the other thing, by means of which the thing that causes motion causes it, has need of something that causes motion by means of itself prior to that,[502] then first among the things that cause motion like this[503] will be that which causes motion by means of itself. And that which causes motion by means of itself is moved, but not by something else. Therefore [it is moved] by itself. Therefore that which is the first mover and the first moved is self-moved. For even if the thing that is moved is not immediately moved by something that is self-moved but [only] through other things serving as intermediaries, it will [still], properly speaking, be moved by the self-moved thing, without which the intermediate thing could not be moved or cause motion at all.

He has shown the necessity of the argument by [the words] 'when- 1224,1
ever, then, one thing after another causes motion by means of

something [else]', etc. For whenever a mover moves another thing through some intermediary, there must then first be something that causes motion by means of itself if we are not to go on placing intermediary before intermediary to infinity. (On this occasion (*nun*) he has used 'by means of something' [solely] of that which causes motion by means of something else, although he earlier said that something that causes motion by means of itself also causes motion by means of something.)[504]

With regard to the matters at hand there are also two puzzles for me to consider. One is how it is that he has said that it is so [very] absurd[505] that there should not be any first or last mover when we too say that when the elements come into being from one another there is nothing that is the first or last cause of motion but the motion cycles around[506] to infinity. After all, in *On Generation*[507] we learnt that water comes from earth and air from water and fire from air and air and earth once more from fire. And, second, one might wonder how it is that he says that everything that is moved is in the first place moved by the self-moved if we say that the elements are moved by one another in a cycle according to the laws of [inanimate] nature and don't move one another in the same way as living things.[508]

Well, perhaps one should respond to both these puzzles that the elements come from one another *qua* material entities one after another[509] and thus their generation bends into a cycle. For air [comes from] water, not because it creates or moves the air, but because the water turns into air. But here (*nun*) the argument is seeking the mover that causes the thing that is moved to be moved as found in the case of things that are moved and cause motion in place, among which there must be some first mover and last thing moved and the first mover [must be] self-moved lest the argument go on claiming to infinity that every mover is itself moved from outside by something else.

This needs to be articulated yet further,[510] but let us look at the next part of the text.

256b3-13 In addition to the above, the same conclusions will result from looking at it this way. [[511]If everything that is moved is moved by something that is [itself] moved, either this is an accidental property of things, so that a thing that is moved causes motion but not because it is moved itself, or it is not but is a *per se* [property of things]. First, then, if it is accidental, it is not necessary that a thing that causes motion should be moved. And if this is so, it is clear that it is possible that at some time nothing that exists should be in motion; for the accidental is not necessary but may not be the case. Now if we assume a thing that is possible is the case, nothing impossible will result – although perhaps something false will. But for there not to be motion is impossible], for it was shown[512] previously that there must always be motion.

This argument has seemed to some,[513] as Alexander says, [properly] to come after the argument after it, the one that shows that part of the self-moved thing is moved and part causes motion while being unmoved [and] whose first words are: 'Yet surely, if there were in fact a need to consider whether what moves itself or what is moved by something else is cause and principle of motion'.[514] For the proof that the first mover is unmoved which ends, 'Further, if it moves itself as a whole, one part of it will cause motion, while another will be moved. Therefore AB will be moved both by itself and by A [alone]'[515] seems to follow that [argument] and what is said here seems to come after [both of] these by virtue of the words 'In addition to the above, [the same conclusions will result] from looking at it this way', etc.; for what is demonstrated by those [arguments] is also established by these.[516] Accordingly, some think that those [arguments] should have been written and should be read before [these].

However, if the present textual (*gegrammenos*) order is observed, which even Alexander judged the sounder [procedure], [the argument] is promising to prove, by means of a different kind of demonstration, 1225,1 the same thing as those that have gone before [namely,] that among the things that are moved and cause motion, the self-moved is the first mover. After throwing in[517] [the statement] that the first mover *tout court* must be unmoved, he attaches the account of the self-moved [mover], the first,[518] not of movers *tout court*, but of those that, as well as being moved, also move themselves,[519] inquiring of what type the motion within it is and showing that that which causes motion in it causes motion while being unmoved [itself]. Or, rather, he does show this, but nevertheless shows along with it that either the unmoved or the self-moved is the principle of motion. And this latter is shown along with its having been shown that not everything that is moved is moved by something from outside which itself too is moved by something from outside. He has already shown this by other arguments, and he shows it again now employing the destructive *topos* from consequence and division.[520]

The argument goes like this. If everything that is moved is moved by something that is itself moved from outside, the thing that is moved causes motion either accidentally or *per se*. The thing that is moved causes motion accidentally when it does not cause motion by being moved [itself] but may be moved or not moved when causing motion. So if, as he will show,[521] not everything that causes motion is also moved itself either *per se* or accidentally by something from outside, it is clear that not everything that is moved is moved by something that is itself also moved from outside, but that, while it is certainly the case that a thing that is moved is moved by something, the first mover is either self-moved or unmoved. And if that which is moved is moved by something from outside that acts as prime mover, the mover is unmoved; but if that which is first moved is not moved by anything from outside, [the mover] will be self-moved.

That it is not possible for that which causes motion to be moved accidentally he shows as follows. If that which causes motion moves the thing that is moved accidentally and not because it is also moved itself,[522] that which causes motion is not moved necessarily. And if [it is] not [moved] necessarily, [then] it is possible for what is moved accidentally and thus causes motion not to be moved at some time. So if we assume that that which can come about has come about (i.e. that what causes motion accidentally is at some time not[523] moved), a falsehood can follow, if it is also the case that what is not moved [itself] never yet caused motion, in which case (*hote*) the hypothesis is also false.[524] It is certainly not impossible for what is not moved to cause motion, since the hypothesis that states that [movers] are moved accidentally is also not impossible, and in fact (*alla*) he assumed that it is possible [for them] not to be moved. But it was shown[525] in the *Analytics* that an impossibility does not follow from something that is possible. So if, when it has been hypothesised that what causes motion[526] is not moved [itself], something impossible follows, [namely,] that there is no motion, the hypothesis of the accidental [mover] is impossible, and from [the hypothesis] that everything that is moved is moved by something that is moved [itself] it follows that there is no motion, if it is assumed that [the mover] is moved accidentally. For, on (*dia*) the hypothesis which states that everything that is moved is moved by something which is moved [itself], if the mover were not moved nor would it cause motion. But if there were not something that caused motion, nor would there be anything that is moved; and if there is nothing that is moved, motion does not survive either, for motion is [located] in the thing that is moved.[527] But it is impossible for there not to be motion; for it has been proved, and is evident, that there is always motion.

Alexander says that the words 'although perhaps something false will' are present (*eirêsthai*) because it is not necessary that if the premiss is false the conclusion should also be false; for it has been shown that something true is inferred even from false [premisses]. But perhaps that is not the reason for the addition of 'perhaps' here, but it is rather because, should it be hypothesised that that which causes motion is moved accidentally, and should it be assumed not to be moved, the assumption is at times false (if it happens to be moving) but not impossible, for what has been hypothesised to be moved accidentally may also at times not be moved.

256b13-27 And this is the reasonable conclusion. [For there must be three things: (1) that which is moved; (2) that which causes motion; (3) that by means of which [a thing] causes motion. (1) That which is moved must be moved but does not necessarily cause motion. (3) That by means of which [a thing] causes motion

> [must] both cause motion and be moved, for it changes along with that which is moved, with which it coincides in time and space[528] (this is clear in the case of things that cause local motion [and the things they move], for they must be in contact with one another up to a certain point. (2) That which causes motion in such a way that it is not that by means of which [something else] is causing motion[529] is unmoved. Since we observe a last thing which can be moved but has no principal of motion, and [also] that which is moved but not by another thing but by itself, it is reasonable, though not necessary,[530] that there should also be a third thing which causes motion while being unmoved. And for this reason Anaxagoras speaks rightly when he says that mind is impassible and unmixed, since he makes it a principle of motion; for it would only cause motion in this way by being unmoved] and only rule by being unmixed.

Having refuted the argument which states that everything that is moved also thereby accidentally causes motion itself, he must [now] turn to showing that nor is it the case that that which causes motion does so by being moved *per se*, but first he demolishes the entire hypothesis [that came] before the division into accidental and *per se*. This hypothesis was the one that stated that everything that causes motion does so by also being moved itself by something else. And he demolishes it by showing that before the things that are moved and thereby cause motion there is something that is unmoved and thereby causes motion. And then he refutes the second section of the division, the one that states that it is by being moved *per se* that everything that causes motion also causes motion itself.

For the present, [then,] he shows, from a general perspective, that not everything that causes motion does so by being moved itself by showing that it is reasonable that the first mover should be unmoved and that there is something which causes motion while being unmoved. In showing this, he assumes that in motion there must be three things: that which is moved (which can be clearly observed); that which causes motion (for everything that is moved is moved by something that causes motion); and, intermediary to both [of these], that by means of which the thing that causes motion does so. For it is posited that everything that causes motion moves something, namely, that which is moved, and [does so] by means of something, namely, the thing by means of which it causes motion. There being these three, then, that which is moved must be moved by that which causes motion but need not also cause motion; for if something is moved, there is no necessity that it should move another thing, since the last thing moved is moved but no longer moves anything else. On the other hand, that by means of which the mover causes motion must both be moved and cause motion. For that which moves some third thing through something [else] would not move

that thing unless it first moved that through which it moves [it]. For example, someone moving a stone with a stick and by means of the stick[531] moves the stick first; for the things between the first movers and the last things moved must both cause motion and be moved. He himself shows the necessity [of this] when he says, 'for it changes along with that which is moved, with which it coincides in time and space'. For a thing that causes motion immediately by doing so through contact must change along with that which is moved by it, as he showed[532] in
1227,1 the previous book. And if it changes along with [it], it is clear that it too is moved. And that not only things which cause local motion immediately are in contact with the things moved by them, but that those [which cause motion] with regard to the other kinds of motion are too, he [also] showed[533] in the previous book; now, however, because they provide the clearer instances, he chooses to consider the matter only in relation to things that cause local motion.

He has added that things that cause motion or are moved locally [must] be in contact with one another 'up to a certain point' because of things that throw or are thrown. The thing that throws must necessarily be in contact with the thing that is thrown and be moved along with it, but it is not possible for it to always be next to [it] or to change along [with it]. For, after throwing and releasing it, it is no longer in contact with it, but [only] the air that carries it is.

There being, then, two things, [namely,] that which is only moved, and that by means of which a mover causes motion, which causes motion by being moved, there must also be something that causes motion not in the way that the thing by means of which a mover causes motion does (which causes motion by being moved), but which only causes motion, being unmoved [itself] – like that which was only moved and did not move anything else. For, given that there is [a share of] both that which is moved and that which causes motion[534] in the middle [thing] which is both moved and causes motion,[535] if that which is moved also exists on its own, so that it no longer causes motion, it is reasonable that what remains, [namely,] that which only causes motion but is no longer also moved, should [also] exist on its own. And it is reasonable that if the middle partakes of both and is composed of both that which causes motion and that which is moved, and one of the simple entities of which it is composed exists on its own, it is reasonable that the other thing in the combination should be simple and exist on its own.[536] For it is reasonable that each of the extremities of which the middle is [composed] should have the same relation (*analogia*) to the middle. For just as, in the case of mead[537] mixed from wine and honey, when the honey is on its own, the wine must be on its own too, so too, in the case of that which causes motion and is moved, is it reasonable that if one [ingredient] is on its own so is the other.

He wrote 'reasonable, though not necessary'[538] because the statement seems to have been inferred from a kind of consequence and not from demonstrative necessity.[539] But perhaps it is also necessary that if one of the ingredients of the mixture is on its own the other is surely (*pou*) also on its own. Unless, perhaps, one were to reflect that, amongst the elements of speech, the vowels are separately pronounced, but those referred to as consonants do not have any sort of pronunciation on their own but when combined with the vowels are pronounced along with them. So Aristotle was writing quite precisely when he said, 'reasonable, though one would not say necessary', because, although it is so in most cases, in a few it is otherwise.

He further supports [the hypothesis] that the first mover is unmoved by means of an argument resting on opinion[540] by commending the view of Anaxagoras, who, after hypothesising that mind is the first mover, hypothesised that it is unmoved and unmixed and simple, on the ground that the first mover can only rule the universe (*ta hola*) if it is unmoved and impassible. For if it is going to cause motion by being moved, [this will be] by being moved either *per se* or accidentally, [but] of these [alternatives being moved] accidentally has turned out to have an impossible consequence, [namely,] the non-existence of motion, while [being moved] *per se* will be shown to be impossible in what follows.

> **256b27-257a3** But if that which causes motion is moved not 1228,1
> accidentally[541] [but necessarily, and if it were not moved would not cause motion, [then] that which causes motion, in so far as it is moved, must either be moved with the same kind of motion[542] [that it causes] or with another [kind]. I mean that either that which heats is heated itself and that which brings about health is made healthy itself and that which carries is carried itself, or [alternatively] that that which brings about health is carried and that which carries is increased. But it is clear that this[543] is impossible. For one must divide all the way down to individuals (*atoma*)[544] and say, for instance, that if someone[545] is teaching geometry he is being taught the same geometry, or if he is throwing, he is being thrown] with the same kind (*tropos*) of throw.

Having criticised (*anteipein*) the entire hypothesis that states that everything that causes motion does so by being moved itself, and having refuted it for the time being both by means of [the consideration] that not everything that is moved accidentally causes motion[546] and by showing, for now[547] (*dia mesou*), that it is reasonable that there should be something that causes motion [while being] unmoved [itself] (he will soon establish[548] this by means of demonstrative necessity), he turns to the second section of the division that states that if everything that causes motion does so by being moved itself, it is moved[549] either accidentally or *per se*. (It has [already] been shown that an impossibility, [namely,] the non-existence of motion, follows upon the 'accidental' alternative.)

He says, 'But if that which causes motion is moved not accidentally but necessarily, so that if it were not moved, it would not cause motion'.[550] By this he shows what *per se*[551] means, [namely,] that it is that which is of necessity such that unless it is moved it does not cause motion. From which it is also clear what kind of thing the accidental is, [namely,] that it is not necessarily the case that it does not cause motion unless it is moved, but it is possible for such a thing to cause motion without being moved.

He shows that it is also not the case that that which causes motion does so by being moved necessarily, that is to say, *per se*, by dividing that which is moved *per se* and thereby causes motion into things which are moved with the same kind of motion that they cause (for instance, 'that which heats [and] is heated itself and that which brings about health [and] is made healthy itself and that which carries [and] is carried itself')[552] and things [which are] not [moved with] the same [kind of motion] but a different one (for instance, 'that which brings about health [and] is carried and that which carries [and] is increased'), and, having shown that both of these are impossible, he is left with one conclusion, that it is also certainly not the case that things that cause motion are moved *per se*.

The [words] 'But it is clear that this is impossible' are primarily directed at someone who says that that which causes motion both causes it and is moved with the same kind [of motion], as is clear from [the continuation] 'for one must divide all the way down to individuals and say', etc., which applies to things which cause motion and are moved with the same kind of motion. But by means of this he will also prove impossibility in the case of that which causes motion with one kind [of motion] and is moved with another, for he will show that this too will come down to that which causes motion and is moved with the same kind [of motion].

He says that the absurdity of both causing motion and being moved with the same kind [of motion] is especially apparent in the case of individuals. For if someone who teaches geometry is learning exactly what he is teaching, and someone who throws is, as he throws, being thrown 'with the same kind of throw', there is an absurd consequence (which he will soon introduce),[553] [namely,] that opposites will co-exist. For if someone who is teaching[554] is at the same time learning the same thing, he both has and does not have knowledge of the same thing at the same time. And someone who throws and is thrown at the same time with the same kind (*eidos*) [of motion] will be both actual and potential[555] at the same time with respect to the same form (*eidos*), since one who throws is active (*energein*) with respect to the form, while one who is thrown is passive[556] (*paskhein*), the one thrower,[557] the other thrown.

He says 'one must divide all the way down to individuals and say '[558]
1229,1 not of individuals properly speaking but of things [grouped] by species

(*eidos*), on the ground that things that both cause motion and are moved with the same kind (*eidos*) [of motion] are divided from a genus.[559] And the examples – 'if someone is teaching geometry, he is being taught the same geometry, or if he is throwing, he is being thrown with the same kind of throw' – are of the same kind, since they[560] are species of motion.

> **257a3-14** And if[561] it is not so but each[562] is from a different genus (*ex allou genous*),[563] [[one will have to say], for instance, that that which carries is being increased, and that which increases it is being altered by something else, and that which alters that is being moved with another kind of motion. But this must come to a halt, since motions are finite [in number]. And to turn back[564] and to say that what causes alteration is carried is the same as saying right away that that which carries is carried and that someone who teaches is taught. (For it is clear that everything that is moved is also moved by a mover higher [in the series], and more so by the earlier of the movers.) But this at any rate is obviously impossible. For the result is that someone who teaches is learning,] and one of them must be without knowledge and the other have it.[565]

Having shown by means of examples the absurdity which results from the hypothesis which states that everything that causes motion is moved with the same kind [of motion] that it causes (*kinein*), and intending shortly to argue[566] further (*epagein*) that the absurdity is general, he has turned to the second section [of the division], the one that states that everything that causes motion is moved, though not with the same motion that it causes (*kinein*) but with a different one. For instance, A might be carried by B, B, while carrying A, might be increased by C, C, while increasing B, might be altered by D, and D might be moved with some other kind of motion by another thing.

He shows that this hypothesis too will [eventually] turn into the previous one, the one that states that everything that causes motion is moved with the same motion that it causes (*kinein*). He presents [his] argument with the support of an axiom to the effect that the species (*eidos*) of motion are finite and not infinite [in number]. For motions existed [as we saw[567]] in three [categories], in place, in quality, in quantity; and in each of these the individual [instances], *qua* species, are finite. The species of local motions, for instance, whether of inanimate or animate things, are not infinite: the places [in which they can occur] are limited, if they are indeed [of] six [kinds], and moreover the motions of living creatures are, in the case of some, [restricted to] flying and swimming, in the case of land animals, [to] walking, creeping, wriggling and the like. Moreover, types of qualitative change are also limited if the qualities are; and they are if they are perceived by the senses and [if] the senses are means of knowing finite things. And so too, of course, are instances of increase, because it was not possible [as we saw[568]] to increase to infinity. Consequently, if these are the only kinds (*eidos*)[569] of motion, D, which moves C, will either (1) no longer be moved, on the basis that that which causes motion must be moved with a different kind of motion from that which it causes (*kinein*) and there cannot be a change to another kind of motion over and above those hypothesised and so it will either cause motion without being moved or not cause motion at all, and if it does not cause motion, nor will things that come after it be moved, and so motion will again be eliminated; or, (2) if it must be moved in order to cause motion, it will be moved with some motion[570] that the things before it were moved with, and will turn back to the same kinds of motion, so that D, which alters C, will need to be carried (if that is what happens [to it]),[571] by some other thing, E. But to say that that which causes alteration is once more carried is tantamount to saying that that which moves something is both causing motion and being moved [with] the same kind of motion. For A, the first thing carried, is moved no more by B, the thing that moves it immediately, than by E, the first thing [to do so], or rather even more by the 1230,1 latter. For it has been shown that the first mover and what is closer to the first mover move the last things [in the chain] more than do the things that move them immediately, since for the latter too the cause of their producing motion[572] stems from the former.[573] It will come to the same thing, then, and that which carries and that which is carried will exist together, if A, which is being carried, is moved by D, which is itself being carried by E.

To the above Aristotle appends [the comment], 'But this at any rate is obviously impossible', because an impossibility results in the first instance for those who hypothesise that both causing motion and being moved [take place] with the same kind [of motion], [both members of this] contradictory pair coming together at the same site (*eis tauto*), and [only] by way of them[574] for those too who say that a thing that causes motion is moved with a different kind [of motion]; for this hypothesis too reverts, as has been shown, to that involving (*kata*) the same [kind of motion].[575]

Next he introduces the impossibility in every case (*koinôs*) consequent upon [the notion of] causing motion and being moved in such a way that the thing that causes motion does so by being moved *per se*, whether it is moved with the same motion or a different one. The impossibility is that whereas that which causes motion is in act when it moves something, that which is moved is in potentiality.[576] So, if [a thing] is mover and moved with respect to the same thing, it will at the same time both have and not have that with respect to which it is hypothesised to both cause motion and be moved, having it to the extent that it causes motion and not having it to the extent that it is moved. For instance, someone who teaches teaches things he knows and that which heats heats when it is hot. So if the latter is at the same time being heated and the former at the same time learning, that which is being heated will not already be hot and that which is learning will not already have knowledge. And consequently they will co-exist in contradiction.

If someone says that something that is being heated can heat, he is not [really] referring to (*lambanein*) the thing that is being heated but to that through whose agency it heats; for that which heats by itself, [and] which also provides the reason for the things in between[577] heating, heats, or teaches, or causes motion of whatever kind (*holôs*), by being actually of that kind; at any rate, without there being something of that kind, none of them[578] would cause motion. Further, even the things that heat in this way[579] cause motion by being moved accidentally: they do not heat by being heated but by possessing heat. It is, at any rate, possible for them to heat even without being heated. And in that case it will [only] be accidentally that they cause motion[580] by being moved. And the contradiction no longer occurs (*suntrekhein*) if it is no longer the hypothesis that everything that is moved is moved by something which is moved [itself]. For even if some things are moved by things which are moved [themselves], at least the first cause for the things that both cause motion and are moved in this manner does not cause motion by being moved.

> **257a14-27** Yet more unreasonable than this [is that the consequence is that, since everything that is moved is moved by something which is moved [itself], everything that is capable of causing motion is capable of being moved: that is, it will be capable of being moved in the sense that one could say that everything capable of producing health is capable of being made healthy and everything capable of building capable of being built, either immediately or through a number[581] of intermediaries (I mean [by this last] as though everything capable of motion could indeed be moved by something else, but could not be moved with the same motion with which it moved its neighbour but [only] with a different one, that which is capable of producing health, for instance, being capable of learning – but as this ascends through the series, it will come sooner or later to the same kind [of motion], as we said earlier). So the one [alternative] is impossible, the other fantastic. For it is absurd that that which is capable of causing qualitative alteration should of necessity be capable of being increased. It is not, therefore, necessary that that which is moved should always be moved by something else which is also moved [by yet another thing]. Therefore [the series] will stop. And so the first thing to be moved will either be moved by something that is at rest] or it will move itself.[582]

Having first shown by way of the types of motion (both the accidental and the *per se*, and in the latter case both with respect to things which

cause motion and are moved with the same kind [of motion] and those that [cause motion and are moved] with different kind[s]) that it is impossible that everything that causes motion should also be itself moved by another thing, he now proves the very same thing more
1231,1 generally. He describes what is proved in this way as appearing '*more* unreasonable' because [arguments] that are proved generally are more cogent than those that are proved individually. For if, he says, everything that causes motion does so by being moved, the result will be that a thing that causes motion is capable of being moved [only] to the extent that it is capable of causing motion and is moved [only] to the extent that it causes motion, and is [thus] capable of causing motion and capable of being moved by virtue of the same potency, which is entirely unreasonable. For if causing motion is doing something and being moved is being acted on and [if] it is impossible for these[583] to exist together in the same thing with respect to the same [activity] at the same time, it will be impossible for everything that causes motion to do so by being moved. And he once more confirms the general [proof] by means of the particular examples on the basis of which he earlier constructed his demonstration. For it is not possible that the same thing should, to the extent that it is capable of producing health, be 'capable of being made healthy, or for that which is capable of building to be capable of being built'.[584] And that is the [logical] consequence for those who say that everything that causes motion does so by being moved [itself]. However it can come about 'either immediately', [as happens] when they say that [a thing] both causes motion and is moved with the same kind [of motion], 'or through a number [of intermediaries]', [as happens] when they say that [a thing] causes motion according to one form[585] (*eidos*) (for instance, causes health according to health),[586] but is moved with another (for example, is altered according to alteration), since for these people too there is again the same result, [namely, having] to say that [a thing] causes motion and is moved with the same [kind of motion]. But it[587] will differ in that, because, as was shown earlier,[588] the species of motion are finite [in number], it will reach this [point] through a number of intermediaries, since the last thing moved is moved to a greater extent by the first mover than by the immediate one. And of this pair, [sc. being moved] with the [same] kind[589] [of motion] and [being moved] with another, he states that 'the one is impossible, the other fantastic'. And, as Alexander says, 'he is not saying that the fantastic [alternative] is not impossible (for both are impossible, since both come down to (*proienai*) the same absurdity), but what he is saying is that one [alternative] is impossible and the other in addition to [being] impossible, is both fantastic and incredible. The one that has both [defects] is to think that that which causes motion does so with one kind of motion but is moved of necessity with another. For to say that a thing that alters something of necessity alters [it] by being increased itself and cannot do so otherwise is an absurd hypothesis.'

Thus Alexander. But perhaps Aristotle made a more precise distinction when he said that [causing motion and being moved] with the same kind [of motion] is, for its part, impossible, since it is impossible for acting and being acted on to coincide at the same site,[590] while to be moved with one [kind of motion] and to cause motion with another would be possible if the species of motion were not finite [in number] and if the first thing moved[591] were not moved to a greater degree by the first mover than by the immediate one.[592] On the other hand, to hypothesise that finite things are infinite is truly fantastic, the vacuous nonsense of someone who fantasises that it is possible to assume one motion after another for ever.

In the same way, after presenting the general demonstration, Aristotle drew (*epagein*) the conclusion that 'it is not, therefore, necessary that that which is moved should be moved by something else which is also moved [by yet another thing]',[593] not saying this because it is not always necessary for that which is moved to be moved by something else (for that which is moved has been shown to be moved by some other 1232,1
thing), nor because it will not be moved by some other thing which is moved itself (for he will show that it is moved by that which is self-moved), but because it is not necessary that that which is moved should be moved by something that is always itself moved by something else. It is for this reason (*dio*) that he next concludes that the prime mover is either unmoved or self-moved, being moved from within (*ex*) itself and not by something else. For even though there are some things that are moved by other things from outside which are also moved themselves, as for instance the stone [that is moved] by the stick that is [in turn] moved by the hand, nevertheless this will not go on to infinity but will come to a halt, with the result that the mover will no longer be moved by another thing but be either unmoved or self-moved.

'Once it was shown that that which causes motion does so neither by always being moved with the same kind [of motion] nor [by being moved] with a different one, it followed', says Alexander, 'that the prime mover is unmoved, since [the possibility] that everything that is moved is moved by something which is moved [itself] was [thereby] eliminated. However, he does not yet assume this, since he has not yet (*proêgoumenôs*) proved it.'

But perhaps, given that it has been shown that that which causes motion does not cause motion by being moved by something else, but not that it is not [moved] by something which is moved *tout court*,[594] it follows that the first mover is not exclusively unmoved but either unmoved or self-moved, as Alexander himself later notes.

Someone,[595] observing the transformation of the elements, might, I imagine, be puzzled as to how it is impossible for something that is moved to be moved by something that is itself moved from outside. After all, when fire comes into being, it does so (*ginesthai*) through the agency of[596] [other] fire which itself takes its origin from [yet] other fire, or through the agency of air which rubs together and ignites.[597] And in fact it has been shown in *On Generation*[598] that the elements are for ever coming into being out of one another in a cycle (*anakukloumena*) so that everything that is moved in the sublunar region is moved by something which is itself also moved from outside, such motion proceeding, always cyclically, to infinity, as does generation, even though the species of motion are finite [in number].

Well, perhaps one should say to this that the argument is not presently seeking the material cause of generation, by which [the elements] come into being from one another, but the efficient cause, by which [they come into being] through each other's agency. In which case (*oun*) fire that comes into being does so through the agency of fire which already exists and not through that of [other fire] which is [also] coming into being and so is not moved by something which is [itself] moved from outside; and fire that is kindled by air rubbing together comes into being accidentally and not *per se*, since everything that produces *per se* produces a product (*ginomenon*) like itself.

257a27-33 Yet surely, if there were in fact a need to consider [whether what moves itself or what is moved by something else is cause and principle of motion, everyone would nominate the former. For what is [a cause] *per se* is always a prior cause to what is itself also[599] [one] *per aliud*.[600] And so], making another start,[601] we must consider whether anything moves itself, and how and in what way.

Having shown that the first mover must be either unmoved or self-moved, since what is in motion is not always moved by something moved externally, in the meantime he now proves, by a syllogism such as the following, that what is self-moved is <more of>[602] a principle of motion than what is moved by something else and thus causes motion: what is self-moved is a cause of motion *per se*, as what is moved by something else and therefore causes motion is *per aliud*. The *per se* cause of motion is prior to and more principal than the *per aliud* cause. And it is clear at the outset that the self-moved is a *per se* cause of motion, since it possesses from itself (*ex heautou*) both the fact of causing motion and of being moved. But what is moved by something else, in order that it may cause motion, is *per aliud*. It is also clear that a *per se* cause of motion is more principal and causal than the *per aliud*. For *per aliud* is said in two senses: either accidentally or because of something else (*di' allo*), but neither does what is by accident exist unless what is *per se* precedes it, nor does what is because of something else [exist] unless what is because of itself (*di' hauto*) – that is, what is *per se* – exists previously. Thus, what is self-moved is the first cause and the principle of the things that cause motion and are moved. Next, investigating how such a thing moves itself, he will discover and prove once again, from its division into what causes motion and what is moved, that the first mover is unmoved. By 'taking another starting-point', he means the one according to which everything in motion is divisible, which he demonstrated in the fifth book of this treatise,[603] but this was not the one he made use of in what was assumed beforehand, but the one that says that everything in motion is moved by something, and the one [that says] that everything in motion is moved either according to nature or contrary to nature. This, then, is why he said 'taking another starting-point'. He investigates how what is self-moved is moved, and what is its mode of motion, discovering on the basis of this that of beings, some are unmoved, others always moved, and others sometimes in motion and sometimes at rest.

> **257a33-257b1** It is necessary, then, that everything that is moved [should be divisible into divisible [parts] to infinity (*eis aei*). For it has been shown previously in the general [observations] *On Nature*] that everything that is moved *per se* is continuous.

In the fifth book of this treatise, he proved that everything in motion *per se* is continuous and divisible. This was proved by the fact that the object in motion is neither in that toward which it is being moved, nor is it any longer in that out of which it is moving, but a part of it is in the latter, part of it is in the former, and such a thing is divisible. He therefore now makes this the starting point of his investigation into what is self-moved. One can also reason from this, that when he says he has said something in the *On Nature*, he means the first five books of this treatise, but when he says it was in the *On Motion*, he would mean the last three.

> **257b2-6** It is impossible, then, for what moves itself [to move itself in its entirety, for it would be transported and transport as a whole by the same [act of] transportation, being one and indivisible in form, or be altered and alter [as a whole] so as to teach and learn at the same time] or cure and be cured with the same health.

Eudemos,[604] having previously recounted that what moves itself was propounded by Plato, first divides the ways in which something may be said to be self-moved, writing as follows: 'for either the whole moves the 1234,1
whole, or the part moves the whole, or *vice versa*, or the part [moves] a part. But we say that causing motion primarily means causing motion by itself, not because of something else, and because of itself, not of something else'. Thus, having eliminated these three modes one by one, he shows that the only thing that can be self-moved is what causes motion by a part and is moved in a part, and of this, what has within itself one aspect that causes motion but is immobile *per se*, and another

that is in motion. Aristotle, however, did not adopt this division, but as he proceeds he does postulate it concisely, when he says 'for if the whole is moved by itself, it will be moved either by one of its parts or as a whole by a whole'.[605] Now, however, discussing each of the modes, he immediately shows the absurd consequences that follow from them. First, having shown by means of examples that the [proposition that] the whole is moved by itself as a whole is incongruous, he next displays that it is impossible. The examples: 'for it would transport as a whole', he says, 'and be transported by the same transportation, being one and indivisible in form, or it would be altered and would alter, so that it would teach and learn at the same time, and would cure and be cured by the same health'.[606] Eudemos, using the same terminology, made this clearer by writing as follows:[607] 'is it possible to say that a whole [person] is carrying a whole thing on his shoulder, and at the same time, one and the same thing is being carried and is on the shoulder, and is carrying something on its shoulder and is holding it? Or perhaps this is not possible, whereas a whole can be heated by a whole? But how? For what heats does so because it is hot, but what is being heated changes toward what is hot. But it is not yet that into which it changes, and it is impossible for the same thing to be hot and not hot at the same time'. 'Aristotle added "and indivisible in form"[608] to "being one" either, as Alexander says, 'since it too is said of what is in motion', just as "what is one", because "in an indivisible form" is equivalent to "one". Alternatively, one of them was said of the object in motion, while "indivisible in form" [was said] of motion, so that the passage that follows should mean: "for it will cause and undergo motion at the same time, by the same transportation, although it is one and indivisible in form".' This being what Alexander says, perhaps it is better to understand both 'being one' and 'indivisible in form' as applying to the entire self-moved thing, while 'indivisible in form' indicates the sense in which 'being one' was said: viz. that it is one not numerically but in form, with 'indivisible' being taken instead of 'one'. For a whole transporting itself and being transported as a whole, and altering and undergoing alteration, is one in form, but not in number, such as this individual chameleon, for instance, that colours itself. I also take issue with how Alexander says that what is in an individual form is one in number. For I agree that 'in an indivisible' means one, but I say that it is one thing to be numerically one and another to be one in form, and Aristotle unambiguously indicates how he meant 'being one': not numerically, but in form. He signified the same thing shortly before, when he said 'for one must speak after dividing down to indivisibles',[609] where here too he means by 'individuals' the ultimate species, not those that are determined numerically. He was right to speak of healing and being healed by the same health, for if what heals were one thing, and what is being healed something else, the health in the one that is active would be different from the health in the one being passive. But since what heals and what is being healed is the same, the health too would be one and the same. For the same [activity] to be active and passive is, however, absurd, as he will demonstrate more clearly next.

> **257b6-13** Further, it has been determined that what is moved is the moveable. [And this is moved in potency, not in entelechy, and what is in potency is advancing into entelechy, and motion is the imperfect entelechy of the moveable. That which causes motion, on the other hand, is already in act. For instance, what is hot heats, and in general what possesses the form engenders [it in other things]. And so at the same time the same thing, and in the same respect, will be hot and not hot. And similarly in each other case where what causes motion must have the synonymous [property].] Therefore one part of what moves itself causes the motion and another is moved.

Having indicated that the argument stating that some whole is moved by the whole of itself is incongruous, he now demonstrates that it is also impossible, as he said previously. He proves it by making use of the definition of motion and of the object in motion. For we call 'in motion' what is moveable [when it is] still in potency and progressing toward complete entelechy, since motion, although it is the entelechy of what is movable, is an imperfect entelechy, because the object in motion is still in potency, whereas what causes motion is already in act (*energeiai*). For what heats must be hot in act, and what gives a share of something to something, which he himself called 'to engender',[610] must itself have beforehand that of which it imparts a share. Since these things are self-evident, if some whole is heated by the whole of itself, insofar as it heats it is already hot in act, but insofar as it is heated and is progressing toward being hot, it possesses this [property] only potentially as long as it is not yet hot, so that 'at the same time the same thing, and in the same respect, will be hot and not hot', which is absurd, since it is impossible for both members of a contradictory pair to be true at the same time. But as things are in the case of what is hot, so they are in the case of the other motions, in which what causes motion has the objects in motion synonymous with itself, as for instance what is transporting makes something be transported, and what increases [makes something] be increased, and what is knowledgeable makes something knowledgeable, the contradictory pair must be true at the same time, if something were moved by itself. But if this is impossible, it is impossible for any whole to be moved by itself as a whole. This impossibility is also concluded in the second figure, as follows. All that is in motion is imperfect, insofar as it is in motion. Nothing that causes motion is imperfect insofar as it causes motion. Therefore, nothing that causes motion, insofar as it causes motion, is moved.

Leading the argument to the case of the hot, Aristotle added 'Similarly in each other case where what causes motion must have what is synonymous',[611] that is, bring what is moved by it into synonymy with itself, as for instance what heats, since it is hot, makes what is being heated hot, and the teacher, being knowledgeable, makes the person who is being taught knowledgeable. He says this because not all things that produce produce things similar to themselves; for a whip (mastix) produces weals, although it does not have weals itself.

1236,1 In the third book of his *Physics* or *On the Heavens*, Theophrastus[612] divides what comes into being as follows: 'Either they are generated by something similar, he says, as man by man and the hot by the hot, or by a contrary, as we see thunder and lightning: the coming into being of this fire in the air is generated by cold, gathering the heat within them into one and igniting it. Third, by what is in entelechy in general, as the weal: for it is generated by a whip that is in entelechy, but is no longer similar or contrary to what is generated. And what is generated by the sun', he says, 'is also generated by an entelechy: for it too is neither similar nor contrary to what is generated by it'. Alexander objects that what results from the whip is said to be generated by an entelechy instead of by actuality and a blow, and it is clear that everything else that comes into being is generated by an entelechy, whether entelechy designates being what it is in act (*energeiai*) or the activity (*energeian*). Perhaps he named them in this way because the other things are dominated by some proper characteristic, while these have only being active (*to energein*). I believe it is worthwhile objecting to the division into things that produce and things that are generated: perhaps everything that produces *per se* and is generated naturally produces similar things and becomes similar. For what comes into being is necessarily generated out of contraries as its matter, but nothing is generated *per se* by contraries as by efficient causes. For contraries have an effect on each other, yet it is not the contrary that is generated by a contrary, but what is similar. This is also the view of Aristotle, who writes as follows in the first book of the *On Generation*:[613] 'It is therefore reasonable that fire heats and what is cold cools, and in general what produces assimilates to itself what is passive. For what produces and what is passive are contraries, and coming-to-be is toward the contrary, so that what is passive must change into what produces'. Fire, too, is generated by what is cold accidentally, and not *per se*: for it is not insofar as it is cold that it produces fire, but insofar as it is cold it densifies and brings together the fires that were previously scattered, which, by rubbing them against one another, it makes ignite even further, changing into themselves what approaches them, and the whip *per se* imparts upon the person being whipped a mark similar to itself, violently impelling it through motion, but the welts supervene accidentally, when numerous small blood-vessels are broken and the blood flows internally. If, then, these things I say are

true, Aristotle seems to have said 'in every case where what causes motion must have what is synonymous in motion' instead of 'whatever causes motion *per se*', indicating this as if from the testimony of what is synonymous, and wishing the contradictory opposition to be also true in the case where someone says that a thing is itself moved by itself,[614] but not in the case of things that cause and undergo motion by accident. Yet having shown that it is impossible for something to be self-moved in the sense that it is moved as a whole by itself as a whole, he also 1237,1
concluded that if there is anything self-moved at all, part of it must cause motion, and another part of it be in motion.

> **257b13-20** But that it is not self-moving in such a way [that each part is moved by the other is clear from the following. First, there will be no first mover if each part will move itself. (The earlier will be more the cause of the motion than what comes after and will cause motion to a greater degree, for it was possible [for a thing] to cause motion it two ways, by being itself moved by something else, or by itself, and] what is further from the thing that is moved is nearer to the principle [of motion] than what comes in between.)

Having concluded 'if there is anything self-moved, it is necessary that one part of it cause motion, and another be moved', he again gave us to understand the division of this as well. For if one thing causes motion and the other is moved, either a part causes motion and the whole is moved, or the whole causes motion while a part is moved, or some part causes motion and [another] part is moved. Eudemos[615] divides this, too, as follows: 'the part causing motion, he says, will either do so while being immobile, or [will be] moved. But if it is immobile, what we want has been proved: that what is self-moved is composed of what is moved and an immobile mover. But if the part causing motion does so while itself being moved, since all that is moved is moved by something, this too will be moved by something. Therefore, what moves it cannot be external to it, for the whole could no longer be said to be moved by itself, but by something else moving the part that moves it. Therefore, it will be moved by some one of the things within what is self-moved. Thus, the part causing motion will be moved either by one of its own parts, or by the thing moved. Again, this mover causes motion either while remaining immobile, and the prime mover will be immobile, or in motion. And again, either by a part of itself or by the object in motion. By using mutual subtraction', says Eudemos, 'we will thus arrive at some point at the prime movers: these were what caused motion not by some [part], but moved wholes by means of wholes. For mutual subtraction[616] cannot be infinite, for they would not move if there were no first one. But it has been proved that it is impossible for a whole to be moved by the whole of itself. Therefore, it is necessary for the prime mover to cause motion while remaining immobile'. This is how Eudemos carries out the division. Aristotle, for his part, first eliminates the [possibility that] some portion causes motion, while some other is in motion, such that the object in motion once again moves what moves it, and the mover is moved by what is moved by it. For thus the result is that each is moved by each: both by what is moved by it and by itself. For if A moves B while B moves A, then A will be moved by itself as well, and more than by B: for what is in motion is moved more by its first mover than by its proximate mover, as was shown earlier.[617] Aristotle refutes this hypothesis, then,[618] by reducing it to the absurdity that there would be no first mover, so that one part of them would cause motion and the other be moved by it.

1238,1 For if it does not move itself as a whole, the prime mover would have to be one thing, and what is moved by it would be another. But he proves that there will be no prime mover in what is self-moved in this way, from the fact that each of the parts will be found to be the first mover. But if all are similarly first, none would be first. That each one is first, he proves from the fact that each one not only moves the other, but itself as well, and it is moved more by itself than by what is moved by it. But what moves to a greater degree is the prime mover. That each one moves itself, he proves by [saying] 'what is earlier [in the series] is more the cause of its being moved than what comes next, and will cause motion to a greater degree'.[619] For if, as has been said, A moves B but B moves A, A will be the prime mover and self-mover, and more so than it is moved by B. Similarly, if B moves A while A moves B, B is once again the first mover and will move itself, and more so than A which is what moves it proximately, so that each one will be first, will move itself, and will be moved as a whole by itself as a whole, which has been proved to be impossible. As proving the fact that A is moved more by itself than by B in things that move one another, he adduced the fact that A causes motion by itself, while B moves A while being moved by A; for what causes motion by itself causes motion to a greater degree than what causes motion while being moved by something else. With regard to 'what is further from the object in motion is nearer to the principle of motion than what is intermediate',[620] he said it instead of 'what is more distant from the ultimate object in motion is more akin and proper with regard to the account of a principle than what is between that thing and what is more distant'. For if the last object in motion is C, what moves it proximately is B, and what is more distant is A, A has more the form of a principle than does B with regard to moving C, being more distant from C than is B. And if A had nothing prior moving it, it would be the principle of motion. Likewise, if there are two, and A moves B but B moves A, then A has more of the nature of a principle with regard to the motion of A than does B, which is more distant from itself. [The same is to be said] in the case of B as well. Thus, all are similarly first, and therefore nothing is first. Alexander says that it has been proved by what has been said that neither of them will be set in motion by something external and other, but by themselves. Yet I think it is better to say that what has been demonstrated is that there is no first mover. To establish this, it was admitted that each of them moves itself as well, so that it might thereby be shown to be more in the nature of a principle, and primary.

> **257b20-3** Further, there is no necessity for what causes motion to be moved other than by itself. [Therefore the other [part] moves it in return [only] accidentally. Hence I have assumed that it is possible for it not to move [it]. Therefore part of it is moved,] but the part that causes motion is unmoved.

[This is] the second argument, proving that what is self-moved could not be such that each of the parts moves another. For it has been shown 1239,1
in what precedes that it is not necessary for everything that causes motion to be moved by something else that is external, but the first mover, if it must be moved, is moved by itself. Thus, if B is supposed to be moved by A, it is true to say that A will not necessarily be moved by something external. For if it ever undergoes reciprocal motion from B, B causes motion accidentally, because it so happened that A is one of those things that is moved by something else, and causes motion in that way. But what causes motion accidentally can also fail to cause motion. Thus, if B is assumed to cause motion, then A will not be moved. Thus, B will be in motion, and A will cause motion while remaining immobile. Therefore, the parts of what is self-moved are not necessarily reciprocally moved by each other. Alexander says that 'another reading is in circulation, which is as follows: "in addition, it is not necessary that what is moved should cause motion" ', and he prefers this reading, for what Aristotle says, 'nevertheless, I have assumed that it is possible for it not to cause motion',[621] follows better, he claims, upon 'it is not necessary for what is moved to cause motion'. But if what was written was 'it is not necessary that what is moved should cause motion', how could he conclude what is inferred: 'unless by itself',[622] unless 'by itself' is construed with the object in motion? But Aristotle seems to wish to prove that they coincide, viz. that neither is A necessarily moved by B, nor does B necessarily move A. Beginning by saying with regard to A that it is not necessary for what causes motion to be moved, he then went on to say about B that it is possible that it may not cause motion.

> **257b23-5** Further, it is not necessary for what causes motion to be moved in return; [on the contrary, it is necessary that either something unmoved or something moved by itself should cause motion], if [as is the case] it is necessary that there always be motion.

The third argument, which proves that the parts of what is self-moved are not moved reciprocally by one another, utilises the same approach: that it is not necessary for everything that causes motion to undergo reciprocal motion, since it has been proved [that] what causes motion primarily does so either while remaining immobile, or while being moved by itself. He reminds us of the reason why this has been demonstrated to be so: for what causes motion primarily must be either immobile or self-moved, 'if it is necessary for there always to be motion',[623] as was shown earlier. For if we suppose one thing to be moved externally by another *ad infinitum*, there will be no motion. For there must be a first mover, but of an infinite [number of things] nothing is first.

> **257b25-6** Further, it would be moved with the motion it causes so that what caused heat would be heated.

It has been shown previously that if something that causes motion *per se* were moved, it would cause motion and be moved by the same form of motion, so that what causes heat would be heated. But this is impossible; for what causes heat does so while already hot in act, while what is being heated is not yet hot in act. Thus, the same thing will be hot and not hot in the same respect, which is impossible. This, then, will result, he says, if one part of what is self-moved causes heat while the other is heated; for since the motion of what is self-moved is one, it must necessarily be hot and not hot, which is impossible. For if the parts were taken to be distinct and not subject to reciprocal motion, there would be nothing surprising if one caused heat while the other was heated, as in the body the liver causes heat and the stomach is heated, and one is active, and the other passive. But if what is self-moved causes and undergoes motion as one, exercising one activity, it is absurd for the contradictory pair to coincide in the case of one thing and in the same respect. But they will coincide, if the parts in what is self-moved, undergoing reciprocal motion, are also forced to move themselves by the same motion by which they move each other. He therefore also added this fourth argument, showing that it is impossible for anything to be self-moved in such a way that the parts move one another reciprocally because it so happens that each of the parts is self-moved, and the contradictory pair coincides in their case. Alexander, for his part, wrote this text as follows: 'that it would cause motion by the same motion by which it is moved', which does not differ in concept.

> **257b26-258a2** But surely nor is it the case that either one part or a number of parts of what moves itself primarily[624] [will move itself individually. For if the whole is moved by itself, it will either be moved by one of its [parts] or the whole [will be moved] by the

> whole. Now, if [it is moved] as a result of some part [of it] being moved by itself, this [part] would be the primary self-mover, for if separated it will move itself, but the whole no longer will. And if, on the other hand, the whole is moved by the whole, these [sc. the parts] would move themselves [only] accidentally. So, if [their self-motion] is not necessary, let it be assumed that they are not moved by themselves.] Of the whole, therefore, part will cause motion while remaining unmoved and part will be moved.

Having shown that whoever speaks of what is self-moved in such a way that its parts move one another is obliged to say that each of the parts of what is self-moved is also self-moved, and having inferred the consequent absurdities by a number of approaches, he adds this point to what has been shown previously in general: viz. that nothing can be self-moved in such a way that one or more of its parts move themselves. For if the whole were self-moved, and had one or more of its parts self-moved, it would either be moved by the self-moved parts within it, or as a whole by a whole. But if the whole is self-moved by virtue of having some self-moved part, that part would be what is primarily self-moved, not the whole. The confirmation of this is that if the part were separated, it would be self-moved, but the whole would no longer be so. If it is moved as a whole by a whole, having one or several self-moved parts within it, it could have been said that it has been shown to be impossible for a whole to be moved by itself as a whole, lest the contradictory pair coincide. Here, however, this is not the issue, but to show that it is impossible for a self-moved whole to have within it one or more self-moved parts. If, therefore, a whole is moved by itself as a whole *per se*, 1241,1
the parts would move themselves accidentally. For as continuous parts are moved accidentally within the whole when it is moved, so the parts of what is self-moved would be moving themselves accidentally. But if accidentally, then not necessarily, so that if they were supposed not to move themselves, nothing impossible would be supposed. I, for my part, think that Aristotle brought the demonstration to this point, showing that it is not necessary for the whole to be self-moved in such a way as to have one or more of its parts self-moved. For if it is moved by its parts, it would not be the whole that is primarily self-moved, but if it is moved as a whole by itself as a whole, it is not necessary that it have one or more unmoved parts within it, which is what had to be shown. Alexander, however, says: 'if the parts that are accidentally self-moved are supposed not to be self-moved, on the one hand nothing impossible will be supposed. On the other hand', he says, 'the whole would no longer be in motion if it were self-moved in this way, since its parts set themselves in motion. But if the supposition also took what has one or more self-moved parts to be moved as a whole by a whole, one must conclude that the impossibility applies to the parts as long as the hypothesis holds'. For the point is to show that it is not necessary that what is

self-moved be so in such a way that it has one or more self-moved parts. If, then, it is in any event necessary that what is self-moved be between what is immobile and what is moved by something else, but it must either be such that it is moved as a whole by itself as a whole, or as a whole by a part, or [as] a part by a whole, or as having one or more self-moved parts within it, or as having one part of it causing motion while immobile and another in motion, but all the others have been eliminated, he rightly concludes by saying,[625] as in the case of a line, 'of the whole, therefore, one part will cause motion while being immobile, and another will be moved; for only in this way can something be self-moved', [viz.] if some part of it were causing motion while immobile, and [another] part were moved by it.

> **258a3-5** Further, if it moves itself as a whole, one part of it will cause motion, while another will be moved. Therefore, AB will be moved by itself and by A.

Alexander seemed to me to explain this text in an original way. For he says its goal is to add another absurdity to the argument that the whole moves itself as a whole in such way that each of its parts would be self-moved. But if this is the absurdity, it is clear that each of the parts, being moved by itself, will also move something else. Thus, the whole will be self-moved, even if its parts, which are self-moved, moved other things. For if the parts did not cooperate with one another with a view to motion, but were moved on their own, the whole would not be
1242,1 self-moved, but each of the parts [would be], while the whole would be a collection of many self-moved entities. If, therefore, the parts also move one another, each of them will at the same time move itself and will be moved by something else, and if the motion of the entire self-moved thing is a single motion, all the parts will clearly be undergoing and causing motion by the same motion, so that each of them will be moved many times by the same motion by itself and by other things, which is absurd: for it is absurd to say that by the same motion by which something is moved from itself, it is also moved by the same [motion] by something else. But if it were moved by itself by one motion, and by the part by another [motion], it would be even more absurd: for it is impossible for what is self-moved to be moved by one motion in this way. 'He himself', he[626] says, 'having stated the argument briefly, left it without further elaboration'. So Alexander; but nowhere, as far as I know, did Aristotle suppose that the parts move one another, but he says that AB is moved by itself and by A. Perhaps, therefore, it is better to understand what is said more straightforwardly: that having said 'of the whole, therefore, one part will cause motion while being immobile, and another will be moved', he confirmed what was said by what follows. For if the whole moves itself, it must be such that one part of it causes motion, while the other is moved. Therefore, the whole of AB is

said to be moved by itself in such a way that A, which is immobile within it, causes motion, with B, that is, being moved. Thus, the whole is said to be self-moved in the sense that it is [made up of] an immobile mover and something in motion. He will also develop this point in what follows.

> **258a5-21** Since some things that cause motion are moved by something else [and some are unmoved, and some things that are moved cause motion and others do not move anything, that which moves itself must consist of [a part] that is unmoved but causes motion and also of [a part] that is moved but does not necessarily cause motion but may or may not. For let A cause motion but be unmoved, and let B be both moved by A and move what we shall call C, and let this last be moved by B but not move anything. (Even though there can sometimes be a number of steps on the way to C, let us assume just one.) Now the whole ABC moves itself. But if I take away C, AB will move itself, A causing the motion and B being moved, but C will not move itself, in fact will not be moved at all. And obviously nor will BC move itself without A, since B causes motion by being moved by something else, not by being moved by some part of itself. Therefore only AB moves itself. Therefore that which moves itself must have [a part] that causes motion but is unmoved and [a part] that is moved but does not necessarily move anything, either with both of these touching one another] or one touching the other.

Alexander says: 'Having shown that what is self-moved has one part that is an immobile mover, and another that is moved by it, he shows by these considerations what the self-moved must be composed of: out of what is immobile but causes motion, and of what is moved by the latter, but does not necessarily cause motion'. When these things are simply said in this way, however, I cannot grasp how they differ. Perhaps, therefore, Aristotle now explains more fully what had previously been stated concisely, first dividing what is active or passive by motion into three. Since there are two of these, what causes motion and what is moved, and what causes motion does so either while immobile or while being moved by something else, and what is moved is so either while causing motion itself or while not moving anything, there will be two extreme terms: what causes motion while remaining immobile, and what is moved but does not move anything else, while intermediate between them and common is what is moved by something else but moves something else. After all, in what causes motion, what causes motion was accompanied by being moved, and in what is in motion, what is in motion was together with causing motion. Having distinguished these three things, then, and having set them out on one line, he shows in which of these the self-moved resides primarily. That the

1243,1 whole of ABC is self-moved is clear, since it contains within it both the fact of moving itself and that of being moved by itself. 'But if I take away C', he says,[627] which was moved but no longer causing motion, the remaining AB will move itself, since A was causing motion within it, and B was what was moved. C, in contrast, will not move itself, nor will it be moved at all, since it is by itself. But neither will BC move itself if A is taken away, for although B causes motion, it does not cause motion by itself, but by the fact that it is moved by something else, viz. A. Among these things, therefore, only AB moves itself. What is self-moved must thus be composed of what causes motion but is immobile, and of what is moved but does not necessarily cause motion, but if it does cause motion, then something else is taken to be what is moved. One might wonder, I think, why he did not rather demonstrate that AC is primarily self-moved, since A only causes motion, no longer being moved itself, while C is only moved, no longer causing motion itself as well; for these were what the self-moved consists in. Perhaps, then, it was because he considered what is self-moved primarily in animals, where what causes motion is the soul and what is moved is the body, which also sometimes causes motion while being moved itself, insofar as it is what proximately enjoys the mover's power, that he stated that the self-moved resides primarily in AB, in which B does not cause motion always or necessarily, but is taken as moved. Nevertheless, the same thing sometimes tends to cause motion itself primarily and to be, in the proper sense, properly dependent on a mover. Yet having said that it is necessary for what is self-moved to have one part that causes motion but is not moved, and [one part that is] 'moved but does not necessarily cause motion, but either does so or not, as the case may be',[628] he adds that what is self-moved possesses these features 'in contact either both with each other, or one with another'.[629] For if, by hypothesis, what causes motion is a body, as is what is moved by it, both will be in contact with each other, for their limits will be together, and those things are in contact whose limits are together.[630] But if what is moved must necessarily be a body, while what causes motion is incorporeal and without size, then both will no longer touch one another, but one will touch another, and not in the proper sense, but metaphorically; for the incorporeal touches the corporeal, as is said in Book 1 of the *On Generation*,[631] where he says: 'Thus, if something causes motion while remaining immobile, it would touch the movable object, but nothing would touch it; for we sometimes say that a man who causes us pain touches us, but we do not touch him'. For what is active could be said to touch more than what is passive. I believe he said this in order to indicate that what causes motion and what is moved would by no means be continuous, but if both are bodies, they would be touching one another in the proper sense, while if one is a body and the other incorporeal, what causes motion, which is an incorporeal, could be metaphorically said to touch what is moved, which is a body. I found the reading as follows:[632] 'and what is moved, on the one hand, but on the other does not necessarily move anything, both either touching one another or one touching the other'. Alexander adds 'by necessity' to what follows, writing thus: 'by necessity either both touching one 1244,1 another or one touching the other'. And each reading can preserve a concept appropriate to the subject at hand, but I think it is more appropriate to construe 'of necessity' with 'not moving anything', since what is moved within what is self-moved, like the body in an animal, often appears to be causing motion itself as well, but not, however, causing motion of necessity. If, however, 'of necessity' were not added to 'in motion but not moving anything', so that it means 'not causing motion by necessity', what is in motion in what is self-moved by the immobile cause will be considered not to move anything at all in any way, which is not true. For since it is proximate to what causes motion primarily, it too receives from it some motive power, as was said previously.

> **258a21-258b4** If, then, what causes motion is continuous [(that which is moved, for its part, *must* be continuous) each will touch the other. It is clear, then, that the whole does not move itself by virtue of some [part] of it being such as to move itself, but that it moves itself as a whole, being both moved and causing motion through some [part] of it being the mover and [some part] the moved. For it does not cause motion as a whole and nor is it moved as a whole, but A causes motion and B alone is moved. There is a problem that arises if one removes anything either from A – if what causes motion but is unmoved is continuous – or from B, [the part] that is moved. Will the remainder of A cause motion or the remainder of B be moved? For, if so, AB would not be *primarily* moved by itself, since when something has been removed from AB, the remainder of AB will still move itself. The answer is that nothing prevents each [of the parts], or one of them, the one that is moved, from being divisible in potency but indivisible in entelechy – and *should* it be divided, from no longer existing with the same nature. And] so nothing prevents [AB, the self-mover,] from residing primarily in things that are [only] potentially divisible.

The text at the beginning of this entire passage, which says that 'if, then, what causes motion is continuous, that which is moved' and so on, seems to Alexander to be incongruous.[633] For he says 'he [Aristotle] spoke in this way as if he were going to give "but if it is not continuous" as the corresponding apodosis, but he fails to provide such an apodosis. Perhaps, then', he says, 'the meaning is something like the following: whether or not what causes motion is continuous still remains unclear. However, since what is in motion within the self-moved is continuous, it is clear that the whole consisting in what causes motion and what is

moved, which is the self-moved, moves itself'. So Alexander. But perhaps, having carried out his demonstration [using the example] of lines, he[634] takes what causes motion to be continuous by hypothesis. Yet the difficulty introduced here also indicates that he supposed what causes motion to be continuous. It says: 'There is a problem that arises if one removes anything either from A, if what causes motion but is unmoved is continuous'.[635] By what is said now, therefore, he wishes to show the mode of motion of what is self-moved, having stated previously that it is composed of what causes motion but is immobile, and of what is moved but does not necessarily cause motion. Corroborating this once again, he says that since what is in motion is necessarily continuous, as was shown previously, and what causes motion being supposed to be continuous, it is clear that it is said to be self-moved, in that all of it moves itself. But something is said to move all of itself not by virtue of the fact that one part of it is such that it moves itself – this is what was previously supposed, but was refuted, for in that case a living being would not be self-moved by virtue of having something self-moved, the soul, within itself – but moving the whole of itself as a whole, and being moved and causing motion as a whole, but not, however, moving itself as a whole and moved as a whole, but by virtue of the fact that one part of it is what causes motion, and another what
1245,1 is moved. For AB does not cause motion as a whole, nor is it moved as a whole, but the only thing that causes motion is A, while only B is moved. For even if B sometimes causes motion, it does not do so necessarily. 'Only' must be understood as common. In some manuscripts, after 'B alone is moved',[636] there is added 'but C is no longer [moved] by A, for that is impossible'. Yet this addition is unknown to Alexander and, as it seems, to most manuscripts. If this reading is ancient, it would mean that what is self-moved is primarily made up of an immobile mover and something moved, which is of such a nature as to sometimes cause motion, being moved itself, but does not, however, necessarily cause motion. That which is only moved, but does not move anything else, such as C (for this is what C was supposed to be at the outset), although it too is moved by A through B, and moved more by A than by B, as was shown previously, nevertheless it is not moved by A in such a way as to complete what is self-moved. For the object in motion that completes what is self-moved is the one that is proximately moved by what is immobile, and also has the capacity of sometimes causing motion.

Having said that what is self-moved is not self-moved because it has a part that moves itself, but because the whole is such, he then raises difficulties against the [hypothesis that] the whole is what is primarily self-moved. 'For if one removes anything either from A', he says,[637] '– if what causes motion but is unmoved is continuous – or from B, [the part] that is moved' or from both, that is, from A and from B, 'will the remainder of A cause motion or the remainder of B be moved', or not?

Either way, something absurd seems to ensue: it is absurd if the remainder does not cause motion or is not moved when a part is taken away, since what is moved was not a part in it, but the whole. Yet he neglected to divide this and to refute it, considering it self-evident. But if the remainder of A does cause motion, or the remainder of B is moved, AB would not be primarily moved by itself, 'since when something has been removed from AB, the remainder of AB, after the subtraction, will still move itself'.[638] He also made use of this argument in the previous book, near the beginning,[639] when he wished to show that what is self-moved is moved by something. For having assumed as obvious that what stands still by virtue of the fact that something else stops moving is moved by something else, and supposing what is primarily self-moved, and removing from it one of its parts (for everything in motion is continuous), and stopping it, and finding, by removing the part, that what is primarily self-moved stands still in motionlessness, he proved that it is moved by something. Here, however, the author of the difficulty, having supposed the remainder is moved, proved that it is not primarily self-moved.

He solves the difficulty by supposing that what causes motion can still be continuous and divisible, since he has not yet proved that it is partless and indivisible. There is, therefore, he says, nothing to prevent either what causes motion from being divisible as well, if this too were supposed to be such, or the object in motion alone. But it is divisible in potency, not in actuality, that is, it is taken not as divided, but as one 1246,1
and continuous. If this is the case, however, then none of the parts is self-moved in act, since the parts do not even exist by themselves in act, so that the whole will be primarily self-moved in act; but nothing prevents the part from existing potentially. When it is taken as already divided in act, by contrast, then what is still in motion after the removal of some part once again becomes self-moved in act. Thus, if the lesser was contained in the first whole taken to be in act, the whole would not be primarily self-moved. If it is not in act, but what is primarily self-moved is predicated of what is in act, then the whole before it is divided would be primarily self-moved. Thus, since divisible things exist potentially, 'nothing prevents' what is self-moved 'from being primarily' in the parts, even if in act it is primarily in the whole. But 'potentially' can be construed with 'divisible', lest someone understand 'divisible' as meaning 'divided'; for in what is potentially divisible none of the parts exists *per se*. But what does not exist *per se* could not be self-moved. Once it is divided, however, it changes its nature and comes to be in act. 'Why, then', says Alexander, 'does he himself, having objected to the <hypothesis that the self-moved is moved from its part>,[640] and saying that the whole is self-moved in this sense, did he not say that the whole would be what is primarily self-moved, but its part, which, even if it is subtracted, remains self-moved?' And he solves the difficulty nicely, [saying] that by saying in this way that the part in

the whole is self-moved, he already supposed it to be moving itself in act. If so, however, it was taken as divided and subsisting by itself. Here, however, the lesser [part], being in the self-moved whole, which is continuous, because what is continuous has not yet been divided, is not moved *per se*, but if it is not moved it could not be said to be primarily self-moved. But he said that 'another nature comes into being, because that [part], being potential until then and not yet existent, came into being in act'. Alexander also tackled this [problem] in another way, [saying] that even if the object in motion is divisible, it will not already be divisible *qua* moved. For if it is divided, it is possible for it to be moved no longer by the mover by which the whole was moved, having come to be of another nature and not remaining such as it was when the immobile was capable of moving it. 'But if', says Alexander, 'the remainder of what is self-moved is moved after the subtraction, the proof in the previous book[641] would not be valid, through which he seemed to prove that everything that is moved is moved by something. Perhaps, however', he says, 'it is not the part that is moved, but since it is in continuity, he supposed it to be at rest'. He said 'let it be divided' not in order that one [part] should be separated from another, but so that, remaining in continuity, it might be taken to be at rest. That he maintained it to be continuous and whole is clear from the things he adduced. For if, he says, CB – which was a part of BA, supposed to be primarily self-moved – is not moved, he says that AB will not be moved, since it still remains one and continuous.

1247,1 For it is ridiculous to say, having subtracted from it CB, that the whole would not move at all, since it no longer existed, if CB had been removed from it. 'This having been said, however', says Alexander, 'we must investigate how the hypothesis is possible that proposes that some part remains at rest in what is primarily self-moved. Someone might say that this is not what he is saying here, but that if a part of CB were removed from the whole of AB, which is self-moved, and stood still, then the whole would no longer be moved, even if what is left behind is moved. For what is left behind is no longer a whole, since a part has been removed from it, but if it neither is nor remains a whole, it could not be moved as a whole. Such a person would say he[642] has argued very logically and dialectically, but not naturally or demonstratively'.

258b4-9 It is evident, then, from these considerations [that the prime mover is unmoved. For whether what is moved but moved by something [else] stops[643] directly at the first unmoved or whether it stops at what is moved but moves and stops itself, in either case the result is that] the first [element][644] in all objects in motion is unmoved.

Having shown that everything that is in motion, whether by nature or by force, is moved by something, and that everything that causes motion does so either itself being moved by something else external, and so on to infinity, or, if this is absurd, it either stops at something unmoved, viz. what causes motion primarily, or at something self-moved, which causes motion while being moved, but is not moved from outside, but by itself, having within itself the cause both of being moved and of stopping. For those things are self-moved that have within themselves the principle both of stopping and of being moved. Yet he has shown that in the self-moved too, what causes motion does so while being unmoved *per se*, while what is self-moved is made up of what causes motion while being unmoved, and of what is moved but does not cause motion by necessity. Having shown these things, then, at length, he now recalls them concisely and concludes on their basis consistently that the result is that the first mover is unmoved in all things in motion, since prior to all things in motion is the self-moved, and in the latter the mover.[645]

Since Aristotle seems to differ in his formulations from Plato with regard to what is self-moved, it would be well to look at this difference as well. Both clearly agree what is self-moved is the principle of motion and of all things in motion, Aristotle[646] saying in this context that 'if one had to investigate whether the cause and principle of motion is what moves itself or what is moved by something else, everyone would suppose it is the former'. Plato,[647] for his part, clearly called the self-moved 'source and principle of motion' in the *Phaedrus*, and in the tenth book of the *Laws*[648] as well, where the Athenian stranger says in conclusion: 'Consequently, as the principle of all motions, the first to occur among bodies at rest and the first in rank in moving bodies, the motion that moves itself we shall pronounce to be necessarily the eldest and mightiest of all changes'. In addition, both say that living beings are self-moved, but Aristotle says that what is primarily and properly self-moved is the animal that is made up of soul and body, of which the body is moved while the soul causes motion while being unmoved by itself, while Plato will have it that what is primarily and properly self-moved is the entire soul, or, as the most exact among the interpreters say, the rational soul. It is through this soul that living beings come to be self-moved, once they partake of a trace of self-motion. And that he would have the soul be what is self-moved in the proper sense, he indicates by giving its definition in the *Phaedrus*, and even more clearly in the tenth book of the *Laws*,[649] when he says: 'Well, what is the definition (*logos*) of the thing for which soul is the name? Can we find any but the phrase we have just used, the substance that can move itself, which has the name we all call soul?'. That he will have it that it is through the fact that the soul is self-moved that the living being is also self-moved, he also indicated in the *Phaedrus*,[650] when he says: 'for any body that has an external source of motion is inanimate, but that

which has it inside itself and from itself is animate, for such is the nature of soul', that is, the one that moves itself and makes what participates [in it] appear to be self-moved. Thus, he made the living being's self-motion a testimony to psychic self-motion, but the cause of the former is the latter. The difference came about insofar as Plato grants that all change, both active and passive, is motion. In the *Timaeus*,[651] the soul 'speaks while in motion throughout its entirety', while in the tenth book of the *Laws*:[652] 'Soul, then, by her own motions leads all things in sky, earth, or sea – and the names of these motions are willing, concerning itself, investigating, deliberating, opining correctly or falsely, joyful or grieving, bold, fearful, hating, loving, and whatever primeval motions are akin to these'. Aristotle, by contrast, who esteems only physical changes worthy to be called motions, considers that the soul is active but is not moved. He clearly denies physical motions of the soul in Book 1 of *On the Soul*,[653] when he says: 'but since being moved is said in two senses' – that is, *per se* and accidentally – 'now', he says, 'let us investigate whether the soul is moved *per se* and has a share in motion. But since there are four motions, translation,
1249,1 alteration, diminution, and increase, it will either be moved with all of them or with several or with all'. Yet Plato also denies them of the soul, enumerating them among the nine motions prior to the soul in the tenth book of the *Laws*, having transmitted the previously mentioned ones as the motions of the soul. And it is clear that Aristotle attributes these to the soul, and knows that they take place from it and toward it, but he calls them not motions but activities and affects. That the soul knows and perfects itself Aristotle necessarily knows, he who has philosophised so much about the soul, but having once assigned to bodies motion's tendency to make things depart from their current state,[654] he quite consistently does not think it well to call the soul self-moved, but the living being, on the ground that only motion in the sense of being moved is evident in the body; and since he also considers this to be the only motion in the thing moved, he sets aside motion [in the soul]. Here, therefore, the difference between the philosophers is not over a reality, but over a name, as it is in most cases. I think the reason often is that Aristotle wishes to preserve linguistic usage and carries out his arguments on basis of what is evident to the senses, while Plato often scorns them and readily resorts to intellective theories. Aristotle, wishing to show in what sense of motion he is here taking 'self-moved', viz. in the sense of bodily motion, in which the soul in unmoved not only according to him, but also to Plato, made the beginning of his demonstration the fact that everything that is in motion is necessarily continuous and divisible into what is always divisible. But it is body that is such primarily. According to this [sense of] motion, then, what would be self-moved in the proper sense of the term if not the living being? For neither can the body cause motion, nor can the soul be moved by this motion, but the body is moved, whereas

the soul causes motion without being moved, since it is unmoved. Therefore, he does not wish any unified whole to be self-moved, because in this sense of motion neither is what is moved able to cause motion primarily, nor can what causes motion be moved. He clearly pointed out the absurdity by saying:[655] 'for, being one and indivisible in form, it would be transported and would transport by the same transportation, and it would be altered and would alter, so that it would teach and learn at the same time, and would heal and be healed by the same health'. Thus, it would be simultaneously in potency and act, simultaneously not yet existing and already being in accordance with what it is said to be. In fact, being something one and simple, it could not be self-moved. After all, the soul, which possesses all the rational formulae (*logous*), has some of them close at hand and active, and others at rest and still in potency within it, and it moves the ones that are still in potency by those that are in act, but the latter are moved, so that the entire [soul] is said to be self-moved. Aristotle also concurs with these concepts, which are Platonic: for Aristotle as well, of the soul, one [part] is the intellect in potency, and another [the intellect] in act, and the one that 1250,1
is in potency comes to be in act by the one that is in act, with the soul teaching itself and learning from itself, and seeking and finding. For all discursive learning, as he himself teaches,[656] that is, both seeking and finding, takes place from the pre-existent knowledge that is in accordance with the intellect that is then active.

This may be why, although he defines self-motion in accordance with natural motions, he interjected among his examples 'it would teach and learn', indicating the same thing. That there exists within the soul both teaching itself and learning from itself, not inasmuch as it is one and simple, which he himself said is indivisible in form, but being divided into what is in potency and what is in act. He therefore led the argument to this absurdity [which states that] the same thing, being one and indivisible in form, would be simultaneously in potency and in act, and both not yet existent and already existent, so that if someone were to take the soul to be divided into potency and act, it would no longer be absurd for the same thing to be in potency and act. The soul is thus self-active, self-knowing, and self-discovering[657] according to Aristotle, yet not *qua* one and simple, but inasmuch as it is made up of potency and act. Yet he did not think it well to call it self-moved, because he observed motion only in bodies; for these are what are continuous and primarily divisible into what is always divisible. This is why motion is also continuous and divisible, and through motion, time. Yet how, according to the definition of motion he laid down, which says that it is the entelechy of what is potential *qua* potential, can the soul fail to be moved and to be self-moved, if it progresses by itself from potency to act? Perhaps this definition, which states that it is the entelechy of what is potential *qua* potential, is common to all change, whether psychic or corporeal, and in general to the change that advances from potency to act; but the definition of motion *qua* motion in the particular sense is 'entelechy of the movable *qua* movable'. Thus, Aristotle does not think it well to call every change a motion, but the one that is habitual in the case of bodies. And not even all of the latter, for he distinguished generation and destruction from motion, because no one would say that what does not yet exist is moved in the proper sense of the term. What is in motion must be something in act, but have something in potency, acting in accordance with which it is said to be moved.

Notes

1. *antilêptikês prothumias*. The expression seems to be a hapax.

2. *telkhines*. Originally a race of Rhodian blacksmith-divinities, they later gained a reputation as evil wizards. According to Eustathius (*Commentary on the Iliad*, vol. II, p. 789, 18 van der Valk), they were proverbially famous as jealous liars and black magicians (*Commentarium in Dionysii periegetae orbis descriptionem*, in K. Müller, *Geographi Graeci minores*, vol. 2, 504, 36), and experts on drugs (*Commentary on the Odyssey*, II, 209 Stallbaum), although, Eustathius adds, some say this may have been slander by those jealous of their ability to work bronze and iron. By extension, the term *telkhines* was subsequently used as a derogatory epithet of sophists and grammarians; cf. *Anthologia Graeca* 11, 321, with the commentary of F. Jacobs, *Anthologia graeca. Commentarius*, II, 2, Lipsiae 1800, p. 177. I know of no evidence to support Wildberg's assumption that *telkhin* could mean 'Christian'.

3. The allusion is to John Philoponus, *Against Aristotle On the Eternity of the World,* written sometime in the 530s. See C. Wildberg 1987, p. 122.

4. i.e. *Physics* 8.1.

5. Aristotle, *Physics* 8.1, 250b13-14.

6. ibid., 250b13-14.

7. *eskhatê tôn zôôn hê phusis*. Cf. Proclus, *in Remp*. 2, p. 311,10 Kroll: 'it is clear that nature is the last life of the body' (*hê de phusis ... sômatos estin eskhatê zôê*); *in Parm*. 961,32 Cousin. Damascius (*in Parm*., vol. I, p. 25,16-17 Westerink-Combès) attributes this view to Aristotle, but the text cited by the editors (*De Part. Anim*. 4,5, 681a12-15) does not seem apt. Cf. Damascius, op. cit., vol. III, p. 35,6-7; p. 50,23-4 Westerink-Combès; Simplicius, *in Phys*., *CAG* 9, p. 287,18ff. Diels.

8. *kai hoion zôn mallon estin ê zôê*. For Proclus, the ontological hierarchy of life moves from pure Life (*zôê monon*), to soul, which is both a life and alive (*kai zôê kai zôn, Elements of Theology* prop. 188), to nature, which is more alive (*zôn*) than it is an active principle of life (*zôê*).

9. Plato, *Laws* 10, 895C, tr. A.E. Taylor.

10. Alexander of Aphrodisias, presumably in his lost commentary on the *Physics*. Cf. P. Moraux 2001, p. 173f.

11. An allusion to the four questions Aristotle identifies at *Posterior Analytics* 2.1, 89b24-5: the fact (*to hoti*), the reason why (*to dioti*), if something is (*ei esti*), what something is (*ti estin*) (tr. J. Barnes). By Late Antiquity, these questions had been developed into a method of scientific enquiry, in which one asked of each subject of investigation, (1) Does it exist? then (2) What is it? then (3) What kind of thing is it? and finally (4) What purpose does it serve? On this investigative scheme, see M. Maróth 1994.

12. Aristotle, *Physics* 8.1, 250b16-17.

13. *hosoi de hena ê mê aei*. This brief clause has been variously interpreted. Ross wished to emend *to hosoi de hena <ê aei> ê mê aei*, while Cornford translated 'those ... who believe that there is a single cosmos or perhaps that there is not always a cosmos in existence'. Simplicius (1122,12ff.), for his part, seems to have read the text as Cornford was to suggest, but understands that the intended contrast is between those who believe (a) there is one world that is always the same, and (b) there is one world, but not always the same.

14. Anaximander, *FdV* 12 A 17, vol. I, p. 86, 20 Diels-Kranz (1951[6]) = Anaximander, fr. 113, Kirk-Raven-Schofield 1983[2], p. 124ff.

15. cf. C.C.W. Taylor 1999, no. 80b, p. 97.

16. cf. H. Usener, 1887, fr. 306, p. 215.

17. *FdV* 13 A 11, vol. I, p. 93, 16 Diels-Kranz = Kirk-Raven-Schofield, op. cit., p. 151.

18. cf. M. Marcovich 1967, no b[6], p. 264 (cited as a testimony to *FdV* 22 B 30).

19. That is, Diogenes of Apollonia. Cf. A. Laks 2008, p. 287.

20. *SVF* II, 576.

21. For this interpretation of *kata tou skhêma tou logou*, cf. J. Bollack 1965, p. 130 n. 5.

22. With Bollack (loc. cit.), I reject Diels' insertion of *aei* at 1121,21.

23. *FdV* 59 A 64, vol. II, p. 21, 36.

24. On Archelaus of Athens, considered the disciple of Anaxagoras and teacher of Socrates, see R. Goulet, *DPhA* I (1989), 333-4.

25. *FdV* 70 A 5, vol. II, p. 231,30.

26. *FdV* 59 B 1, vol. II, p. 32,11. Cf. Simplicius, above, 34,20; 155,23ff. Diels, etc.

27. Plato, *Timaeus*, 30A.

28. This was Porphyry's interpretation; cf. Proclus, *in Tim.* I, 382,26ff. Diehl with the commentary of M. Baltes 1976, I, 146f.

29. Plato, *Statesman* 273D-E.

30. Reading *hupetithonto* instead of *hupethento* at 1122,14 (Diels, in app. crit.).

31. *hosoi de ê hena ê mê aei*. At *Physics* 250b21-2, Ross prefers to read *hosoi d'hena ê aei ê mê aei*, 'those [who say] it is one, whether always or not always'.

32. Empedocles, fr. 17 Diels-Kranz = 8, p. 97,9-13 Wright = vol. 2, p. 43 Bollack.

33. Cornford: 'we must understand him to be speaking of the reversals from the one process to another'.

34. *FdV* 59 A 45, vol. II, p. 18,23-5.

35. *FdV* 31 B 29, vol. I, p. 324, 16f. = fr. 82, op. cit., vol. 2, p. 39 Bollack. Cf. B. Inwood, 1992, 94f.

36. *sphairon eên*. The Greek word for 'sphere' is normally the feminine noun *sphaira*, but Empedocles at least once coins the neuter form *sphairon*. Cf. Inwood p. 94 n. 26; Bollack fr. 83, with commentary vol. 3, pp. 132-3.

37. *kosmos*, one of whose meanings is precisely 'order'.

38. Empedocles, *FdV* 31 B 20 = fr. 60 Bollack = fr. 38 Inwood.

39. *kumbais*. It does not seem possible to identify this bird, cf. Bollack, vol. 3, 106-7.

40. Empedocles, *FdV* 31 B 17, vol. 1, p. 316, 5 = fr. 31, vol. 2, p. 19,10 Bollack = fr. 25 Inwood.

41. ibid., lines 12-13.

42. Reading *apodeixis* with MS F, instead of Diels' *apostasis*, which I cannot construe.

43. cf. Themistius, *Paraphrase of the Physics* 209,19-20 Schenkl (*CAG* 5.2).

44. Aristotle, *Physics* 3.1, 210a10ff. For a summary and discussion of this text, see M. Rashed, 2008, 29ff.

45. Aristotle, *Physics* 8.1, 251a10-11.

46. ibid., 11-14.

47. ibid., 14.

48. ibid., 16.

49. viz. the thing that was at rest but is now in motion (Ross).

50. Aristotle, *Physics* 8.1, 251a16-17, although Simplicius' quotation contains important differences from Aristotle's text, particularly reading *agenêta* instead of Aristotle's *aidia* at line 17 (neither Ross nor Kalbfleisch notes this discrepancy).

51. Aristotle, *Physics* 8.1, 251a20-1.

52. Aristotle, *Metaph.* 11, 1068a1f.

53. Aristotle, *Physics* 8.1, 251a26.

54. Aristotle, *Physics* 8.1, 251a26-7; cf. *Physics* 5.6, 249b25-6; *Physics* 8.8, 264a27-8. Simplicius also takes Aristotle to imply this view at *Physics* 4.12, 221b7ff.; cf. *in Phys.* 742,23f.

55. Aristotle, *Physics* 8.1, 251a27-8.

56. Simplicius also summons Heracles to help him clean the dung-filled Aegaean stables of Philoponus' arguments in his commentary on the *De Caelo*; cf. p. 119,12; 136,1 Heiberg, and the discussion by Ph. Hoffmann 1987, 200f.

57. Philoponus, *Against Aristotle*, fr. 108 Wildberg.

58. Aristotle, *Physics* 3.1, 201b11-12.

59. Aristotle, *Physics* 3.1, 201a19-21 (Simplicius' text differs slightly from that of Aristotle as printed by Ross).

60. Aristotle, *Physics* 3.1, 201b6-7 = *Metaph.* 11, 1065a34-5.

61. Themistius, *Paraphrasis in Phys.* 71,25-72,2 Schenkl.

62. *kata tên aph' henos proodon*. On the meaning of *aph' henos kai pros hen*, 'from one thing and with regard to one thing', otherwise known as Aristotelian 'focal meaning', and its translation into Neoplatonic ontology, see for instance Simplicius, *in Cat.* 32,4ff.; 62,10ff. with the notes to my translation.

63. cf. Porphyry, *Isagoge*, 17,7 Busse.

64. Philoponus, *Against Aristotle*, fr. 109 Wildberg.

65. Reading *<ouk> epharmosei*, following Wildberg.

66. Philoponus, *Against Aristotle*, fr. 110* Wildberg.

67. ibid., fr. 111 Wildberg.

68. Philoponus, *Against Aristotle*, fr. 112 Wildberg.

69. That is, the definition (*Physics* 3.1, 202a7-8, repeated at 8.1, 251a9) that motion is the entelechy or actuality of the movable.

70. cf. above, 1133,21-33.

71. Aristotle, *Physics* 5.1, 224b8ff.

72. Aristotle, *Physics* 3.1, 201a9-15.

73. Above, 1134,17ff.

74. Following the suggestion of an anonymous reader, I replace Diels' comma at the end of 1128,2 by a full stop.

75. ibid., 201a20-1.

76. Reading *energeiai* (with iota subscript) instead of Diels' *energeia*.

77. Although Zahlfleisch (1902, 191) assumes the *autos* of 1138,26 refers to Aristotle, I believe (*pace* my anonymous reader) it refers to Philoponus, and more precisely to his commentary on *On Generation and Corruption*, 120,24ff. Vitelli (*CAG* 14,2), where bread or meat are said to be potentially flesh.

78. On Simplicius' triple division of potentially, see J. Zahlfleisch 1902, 191ff; J. Puig Montada 1999, 150f., and on the distinction between perfect motion (*teleia kinêsis*), equated with actuality (*energeia*), and imperfect actuality (*atelês energeia*), or motion, A. Hasnawi 1994. Cf. Aristotle, *Phys.* 5.4, 228b11-15.

79. Diels' reading *teleiotata tôn allôn* is grammatically difficult and would constitute a hapax. I therefore suggest *teleiotera tôn allôn*. Cf. Philoponus, *in Phys.* *CAG* 17, 905,13 Vitelli (commenting on Aristotle, *Physics* 261b27), where precisely local motion is said to be primary and more perfect than the other motions (*prôtistê kai teleiotera tôn allôn*). For Simplicius, the perfection of local motion (as opposed to the other types of motion, such as generation, alteration, increase/decrease) consists in the fact that local motion does not entail a change in the moved object.

80. cf. Themistius, *Paraphr. in Phys.* (*CAG* 5.2), 69,6ff. Schenkl, tacitly reproduced by Philoponous, *in Phys.* (*CAG* 16), 3.1, 342,16f. Vitelli.

81. Philoponus, *Against Aristotle*, fr. 113 Wildberg.

82. On Simplicius' description of Philoponus as a jaybird (*koloios*), cf. Ph. Hoffmann 1987, 188 and n. 18.

83. Philoponus, *Against Aristotle*, fr. 114 Wildberg.

84. Aristotle, *Physics* 1.8, 191a24ff.

85. ibid., 191b13ff.

86. cf. Philoponus, *Against Proclus on the Eternity of the World*, 9,8, 338,21ff. Rabe.

87. Philoponus, *Against Aristotle*, fr. 115 Wildberg.

88. cf. Philoponus, *Against Proclus on the Eternity of the World*, 9,11, 344,27ff. Rabe. As Diels already noted, Simplicius' reference to 'Book 11' is a mistake, confirming his affirmation that he has not read the book. This section of Philoponus' work was translated (with omissons and supplements) into Arabic under the title *Treatise by Alexander of Aphrodisias refuting the doctrine that one thing can only be generated from another being, and establishing that all things can only be generated out of nothing*; cf. A. Hasnawi 1994, 76ff., with French translation 97-100.

89. Philoponus, *Against Aristotle*, fr. 116 Wildberg.

90. I modify Diels' punctuation here, restoring the missing quotation marks, since Simplicius is here rehearsing Philoponus' argument at 1142,8-14 above.

91. Aristotle, *Physics* 1.8, 191a24ff., cited above at 1140,20ff.

92. See below, note to 1145,22.

93. Simplicius presumably means by this that the doctrine that everything other than the First principle is created (*poioumenos*) is not even consistent with Christian orthodoxy. Indeed, the Nicene Creed states that Christ is 'begotten, not created' (*gennêthenta ou poiêthenta*).

94. The narrow strait separating Euboea and Boeotia, reputed to be winding and therefore unstable.

95. *tên gigantikên thrasutêta*. An allusion to the gigantomachy, the attack of the Giants on Olympus.

96. See above, n. 84.

97. cf. Simplicius, *in De Caelo* (*CAG* 7), 103,19f.; 309,26f.

98. On this principle in Philoponus, see L. Judson 1987.

99. cf. above, 1142,8-14 = fr. 116 Wildberg.

100. Above, 1142,24-6.

101. Aristotle, *Physics* 8.1, 251b5-10.

102. Fr. 117 Wildberg.

103. Aristotle, *Physics* 8.1, 251a16-22; 251b6-10 (with abbreviations and intercalations).

104. I supply *<epi> tôn pros ti* at line 12, following the suggestion of an anonymous reader.

105. Fr. 118 Wildberg.

106. Aristotle, *Physics* 3.2, 202a10-12.

107. Aristotle, *Physics* 8.4, 255b29-31.

108. I translate Diels' conjecture *kai tauta tôn pros ti*.

109. Philoponus, *Against Aristotle*, fr. 119 Wildberg.

110. Reading, following the suggestion of an anonymous reader, *phthorai* (with iota subscript) instead of Diels' *phthora* at line 33.

111. Reading *monou*, with F, instead of Diels' *ekeinou* at line 7.

112. ibid., fr. 120.
113. Reading, at lines 24-5, *anankê prohuparkhein kai ekeino* (MS F) instead of Diels' *anankê prohuparkhein, kai ekeinou.*
114. *kai toutôn phthartôn.* An anonymous reader suggests the translation 'which, moreover, are perishable'. But I take it Simplicius' point is that Philoponus' God is the creator *only of the sublunar elements* that change into one another, and not of the fifth element which, according to Aristotle, makes up the heavenly bodies.
115. Plato, *Timaeus* 41D1-3.
116. Plato.
117. I read *ho de khronos ȩ̂* (*hê* Diels) *kinêsis estin ê arithmos kinêseôs.*
118. Deleting Diels' comma after *kinêsis* at the end of 1153,12.
119. Democritus, *FdV* 68 A 71, vol. 2, 102,3-5 D.K.
120. See above, note to 1121,5.
121. *Timaeus* 28B; 38B.
122. On the links between *aiônios* and *authupostatos*, cf. Proclus, *Elements of Theology*, props. 51-2 ; *in Tim.* I, 279,15ff. Diehl.
123. Plato, *Timaeus* 27D6-28A1.
124. ibid., 28B6-7.
125. ibid., 28B7-C2.
126. ibid., 38B6.
127. ibid., 37C6-D7. Simplicius cites the same text, with the very same introductory formula, above at vol. 9, 703,29ff. Diels.
128. ibid., 38B8-C2 (with slight departures from Plato's text).
129. An etymological play on words: the word 'eternity' (*aiôn*) is conceived as deriving from 'always' (*aei*) and 'existent' (*ôn*). Cf. Aristotle, *On the Heavens* 279a25ff. Simplicius' teacher Damascius (*On the Principles*, vol. I, 218,1 Ruelle) mentions a Phoenician divinity 'cosmic eternity' (*aiôn kosmikos*), so called because 'he has brought all things together within himself' (*hôs panta en heautôi sunêirêkota*).
130. *ekmêruontai.* Another Damascian term (*De Princ.* I, 4,23; 141,25; 158,7; 164,15; 214,17; 282,23; *in Parm.* 89,5-13; 151,28); the image is of a thread unwinding from its skein. Cf. especially a fragment from Damascius' lost treatise *On Time, Space, and Number*, quoted by Simplicius in his *Corollarium de tempore* (*in Phys.* 9, 780,30 Diels), where flowing time is said to unwind itself (*ekmêruetai*) from integral time.
131. Philoponus, *Against Aristotle*, fr. 121 Wildberg.
132. cf. Philoponus, *Aet. Mundi* V, 3, 110,22ff. Rabe. Time is at the third remove from realities in that it presupposes both bodies and motion, being nothing more that the 'number of the motion of bodies' (ibid., 110,17-18).
133. As noted by Wildberg, Diels' *peri tou nun* at 1157,13 is clearly a mistake for *peri tou nou.*
134. I propose to delete *ei* at line 13.
135. cf. Philoponus, *Aet. Mundi* 16, 4, 575,5ff. Rabe.
136. cf. ibid., 576,15ff. Rabe.
137. cf. ibid., 578,15ff. Rabe.
138. In the *Aet. Mundi* (579,14ff. Rabe), Philoponus specifies that this phenomenon is not only theoretically necessary on the hypothesis of the world's perpetuity, but had occurred in his lifetime, specifically year 245 of the Diocletian epoch (i.e. 529 AD).
139. cf. ibid., 580,12ff. Rabe.
140. cf. ibid., 582,24ff. Rabe.
141. Fr. 122* Wildberg. Cf. Philoponus, *Aet. Mundi* V, 3, 107,18ff. Rabe.
142. Diels' *ho ho khronos* (lines 13-14) is, of course, a typographical error.

143. Another trace of a Damascian doctrine; see, once again, the fragments of Damascius' treatise *On Time, Space, and Number*, as preserved in Simplicius' *Corollarium de tempore*.

144. Plato, *Timaeus* 37C6f.

145. cf. above, 1155,1ff.

146. Diels signals a lacuna here. In the preceding paragraph, Simplicius' argument runs as follows: eternity is to the model as time is to the world. But if the model participates in eternity (as it must, since it is eternal) then eternity is higher than (and therefore ontologcally prior to) the model. By analogy, therefore, if the world participates in time (as it must, since it is everlasting), then time is higher than, and hence ontologically prior to, the world.

147. Reading *pantos* at line 29 with MS F, instead of Diels' *pantôs*. Compare Philoponus, *de Aet. Mundi* 574,27 Rabe.

148. A reference to Aesop's fable (Fable 59 Syntipas = 117 Perry) in which a camel asks Zeus for horns and is deprived of his ears instead; cf. Aelius Theon, *Progymnasmata* 75,8-16. As the camel was not satisfied with what he had, so Philoponus is not content with being a grammarian, but wants to be a philosopher as well.

149. cf. Aristotle, *Categories* 12, 14a26-b23. 'Slightly later' (*met' oligon*) is probably a reference to the fact that this chapter was considered to belong to the *post-praedicamenta*, a series of definitional clarifications which the Commentators held followed the treatise on the *Categories* properly so called.

150. Above, 1157,31.

151. Above, 1157,36-7.

152. On the idea that the intellect thinks without transition (*ametabatôs*), cf. Proclus, *Elements of Theology*, props. 52; 211; *in Tim.* I, 248,3ff.

153. Aristotle, *On the Soul* 3.5, 430a19.

154. ibid., 3.7, 431b17.

155. Above, 1158,1.

156. cf. above, 1158,32-5.

157. *pote* is, of course, the name for the sixth Aristotelian category (*Categories* 2a2), sometimes translated as 'time' (as for instance by Cooke in the Loeb edition).

158. Philoponus, *Against Aristotle*, fr. 123 Wildberg.

159. Aristotle, *On the Heavens*, 1, 10, 279b12ff.

160. Philoponus, *Against Aristotle*, fr. 124 Wildberg.

161. For this distinction of *aidiotês* into *aiônios* and *khronikê*, foreshadowed in Plotinus (*Ennead* III 7, 3) and Porphyry (*in Tim.*, Book II, fr. 46 Sodano), see especially Proclus, *Elements of Theology*, prop. 55, 52,30ff. Dodds.

162. Plato, *Timaeus* 28B7.

163. There is no mention of the dodecahedron in the passage of the *Timaeus* (55C) where Plato speaks of the mysterious fifth figure, but cf. Aristotle, *On the Heavens* 307a16; Timaeus Locrus, §35, vol. I, 136,20ff. Marg.

164. cf. Simplicius, *in De Caelo* 12,22ff.; 87,15ff. Heiberg.

165. Xenocrates fr. 53 Heinze = frr. 264-6 Isnardi Parente.

166. *hôs ap' aitias*, 'having an external cause; lacking autonomy' was the last of the four meanings of *genêton* distinguished by Calvinius Taurus (cf. M. Baltes 1976, 108). It was conditionally accepted by Porphyry, as cited by Philoponus, *De Aet. Mundi* VI.17, 172,5ff. Rabe; cf. Baltes, ibid., 143f.

167. That is, the Porphyrian interpretation according to which Plato (*Timaeus* 30A) spoke of disorderly motion in the world prior to Demiurge's intervention only hypothetically, to show what the world would be like if it were bereft of soul (which of course is never actually the case). See above, 1122,6ff. Diels.

168. Above, 1164,25.

169. See, for instance, the opening chapter of the *De Generatione et Corruptione*.

170. Aristotle, *Physics* 8.1, 251b14.

171. Genesis 1:1-5.

172. Simplicius may be thinking of the Hermopolitan ogdoad, whose four male divinities Nun, Keku, Hehu and Amun have been said to correspond respectively to the four primordial elements in Genesis 1: *tehôm* (primeval sea), *ḥôšekh* (darkness), *tôhû wâvôhû* (trackless waste, emptiness, chaos), and *ru^{a}ḥ 'elôhîm* (spirit of God, mighty wind). See J. Hoffmeier 1983, 42ff. Iamblichus already seems to be familiar with this Egyptian cosmology; cf. *On the Mysteries* 8.3, 264,5f. Parthey.

173. Aristotle, *Physics* 8.1, 251b19-23; 26-8.

174. Philoponus, *Against Aristotle*, fr. 125 Wildberg.

175. Aristotle, *Prior Analytics* 2.16, 65a26-7.

176. *SVF* III, 80.

177. On essential attributes (*ta kath' hauta huparkhonta*) as the basis for demonstrations, see for instance Aristotle, *Posterior Analytics* 1.5, 75a27ff.

178. Reading *eikhe* instead of *hexei* at line 27, following Diels' suggestion (in app. crit.).

179. On the completed present tenses (*suntelikoi enestôteis khronoi*), cf. *Scholia on Dionysius of Thrace*, p. 251 Hilgard.

180. Philoponus, *Against Aristotle*, fr. 126.

181. *hupothesin oikeian perainein*. Proclus (*in Tim.* 3,65,1ff. Diehl) accuses Porphyry (*in Tim.*, fr. 79 Sodano) and Theodorus (Test. 17 Deuse) of the same logical sin.

182. Aristotle, *Physics* 8.1, 251b31-2.

183. Alexander of Aphrodisias. Cf. P. Moraux 2001, 174.

184. I punctuate Diels' text as follows: *kai ei men phtharta, kan phthareiê, pasa phthora dia kinêseôs <sc. estin> ktl.*

185. ibid., 32-3.

186. Reading *alla* instead of Diels' *allo* (presumably a typo) at line 23.

187. ibid., 252a1.

188. ibid., 252a2-3.

189. Reading *hôste kai* for Diels' *hôste tai*, an obvious typographical error.

190. cf. Aristotle, *Physics* 3.5, 205a6; 5.1, 224b29, etc.

191. Aristotle, *Categories* 6, 6a18; 11, 14a15, etc.

192. cf. Aristotle, *Physics* 2.2, 194b13.

193. On the distinction between *maransis* and *sbesis*, the two ways in which fire ceases to exist, cf. Aristotle, *On Youth* 5, 469b22ff. The former is self-caused, the latter takes place through contraries.

194. Reading *marainomenês* for Diels' typographical error *manainomenês* at line 11.

195. On this implication in Proclus and Philoponus, see L. Judson 1987.

196. Philoponus, *Against Aristotle*, fr. 127 Wildberg.

197. On the *dia triôn agôgê*, or inferences involving three conditionals ('If, A, B; if B, C; therefore, if A,C'), also known as 'totally hypothetical', cf. Simplicius, above, 759, 14ff. Diels; Theophrastus apud Alexander, *On Prior Analytics I*, 265,16; 326,30ff. Wallies.

198. For the metaphor implied by the verb *perikrouein*, 'to strike a vessel on all sides to see if it is cracked' (LSJ s.v. 2), cf. Plato, *Philebus* 55C.

199. Philoponus, *Against Aristotle*, fr. 128 Wildberg.

200. Aristotle, *Physics* 8.1, 251b31-2.

201. ibid., 251b32-3.

202. Aristotle, *Physics* 8.1, 251b29-31. As Zahlfleisch already noted (1902, 206),

Diels indicates the quotation from Aristotle extends as far as line 37, but in fact it ends at line 33.

203. Philoponus, *Against Aristotle*, fr. 129 Wildberg.

204. For Aristotle's views on motions exempt from the process of generation, see, with Wildberg, R. Sorabji 1983, 11 n. 5. See also Theophrastus, fr. 155C Fortenbaugh et al., with the commentary of R. Sharples 1998, 77-9; M. Rashed 2008, 50. As pointed out by A. Hasnawi 1994, 69f., Philoponus elsewhere cites light, sense perception, and intellection as examples of perfect *energeia* (as opposed to the imperfect *energeia* that is motion); cf. *Aet Mundi* 4, 4, p. 65, 11ff. Rabe; *in de An.* II, 5, 296,21ff. Hayduck (*CAG* 15).

205. Aristotle, *Physics* 8.1, 252a2-3.

206. Aristotle, *Physics* 5.1, 225a1-2.

207. Examples in the commentary tradition of instantaneous change include what we would now call phase transitions (the freezing of water, the curdling of milk or cheese, etc.). See Themistius, *in Phys.* 6.4, 191,30f., with the notes of R.B. Todd (2008, 116-17).

208. *sunêirêmenê*. For Proclus and probably already for Iamblichus and Syrianus, time unfolds (*anellitein*) the multiplicity that is concentrated on the level of eternity (*sunêirêmenon en tois aiôsi*); cf. Proclus, *in Parm.* col. 1235,21-2 Cousin; Simplicius, *in Cat.* 356,26f. Kalbfleisch; *in Phys.* 794,35 Diels.

209. Aristotle emphasised at *Physics* 6, 231b29-231b2, that nothing can be *in motion* and *have moved* at the same instant.

210. Aristotle, *Physics* 8.1, 252a2-3.

211. Aristotle, *Physics* 8.1, 251b29-31.

212. I have altered Diels' punctuation, removing his question mark at line 22 and inserting one at line 25.

213. Reading, at 1174,35, *ton kosmon* with MS F, instead of Diels' difficult *einai*.

214. Following Diels' conjecture in his critical apparatus, I read *hoti* instead of *to*.

215. Philoponus, *Against Aristotle*, fr. 130 Wildberg.

216. Aristotle, *Physics* 8.1, 252a1-2.

217. cf. P. Moraux 2001, 174.

218. That is, as Wildberg points out, in *Against Aristotle* IV.

219. On the sorites or heap argument, see for instance D. Hyde, 'Sorites Paradox', *Stanford Encyclopedia of Philosophy*, http://plato.stanford.edu/entries/sorites-paradox/.

220. cf. above, 1140,20ff.

221. cf. *Physics* 225a2ff.

222. A basic Aristotelian axiom; cf. *Categories* 3b24.

223. See above, 1175,20f.

224. *telesiourgôs* (...) *kai zôtikôs*. In his commentary on *On the Heavens* (373,15ff. Heiberg), Simplicius uses this pair of terms to describe the action of sun's rays, as transmitted by the sphere of the moon to the sublunar world.

225. On the *para mikron erôtêsis*, cf. Sextus Empiricus, *Adv. Math.* 1,69; Galen, *Med. Exp.* 16,2, and in general M.F. Burnyeat 1982.

226. Philoponus, *Against Aristotle*, fr. 121 Wildberg.

227. Aristotle, *Physics* 8.1, 252a1-2.

228. Philoponus, *Against Aristotle*, fr. 132 Wildberg.

229. The final books of the *Against Aristotle*, which have not survived, seem to have been devoted to a description of the world as transfigured following the Second Coming; cf. C. Wildberg 1987, 198ff.

230. On these arguments, cf. H.A. Davidson 1987, 88ff.

231. Aristotle, *On Generation and Corruption* 2.5, 332b30ff.

232. Reading, with Wildberg *hoti *hou* pro tês geneseôs* etc. at line 32 (*ou* Diels).

233. According to Pseudo-Ammonius (*in Anal. Pr.* 68,28f. Wallies [*CAG* 4.6]), the *sun antithesei antistrophê* was the second mode of the Stoic hypothetical syllogism, also called 'the argument that denies the consequent by denial of the antecedent'.

234. cf. Philoponus, *Against Aristotle*, fr. 4 Wildberg (= Simplicius, *in De Caelo* 28,1ff. Heiberg), with Wildberg's n. 4.

235. I omit the *mê* at line 11.

236. The orbits of Saturn and Jupiter respectively.

237. Reading *prosêgage* at line 5, instead of Diels' *proêgage* (cf. below, line 19).

238. Aristotle, *Physics* 3.5, 206a27-33.

239. Time, as a series of moments, never subsists as a whole, but only one moment at a time.

240. cf. Aristotle, *On Generation and Corruption* 2.6, 333a7f.

241. cf. Boethius, *De Topicis Differentiis*, PL 64, 1193Af.

242. *akinêtou tês kinêseôs dedeigmenês.* Since it is not clear in what sense Aristotle has proven motion to be immobile or unmoved, it is tempting to emend *akinêtou* to *agenêtou* : '... since motion has been proven to be *ungenerated*'.

243. Aristotle, *Physics* 8.1, 252a7-9.

244. Eudemos, fr. 110 Wehrli (*Die Schule des Aristoteles*, 82, Basel-Stuttgart 1969, p. 47f.).

245. Empedocles, *FdV* 31 B 27, vol. I, p. 323f. Diels-Kranz = fr. 31-33 Inwood = fr. 92 Bollack.

246. *FdV* 31 B 31, vol. I, p. 325,15 Diels-Kranz = fr. 36 Inwood = fr. 121 Bollack.

247. *FdV* 31 B 17, vol. I, p. 317,10 Diels-Kranz = fr. 25 Inwood = fr. 31, 28 Bollack.

248. *FdV* 31 B 115, vol. I, p. 357,15-16 Diels-Kranz = fr. 11 Inwood = fr. 110 Bollack.

249. Maintaining Diels' reading *erephthê.*

250. *FdV* 31 B 30, vol. I, p. 325,10-12 Diels-Kranz = fr. 35, p. 234 Inwood = fr. 126 Bollack.

251. Reading *ê* instead of Diels' *kai* at 1184,32.

252. Eudemus fr. 111 Wehrli = *FdV* 59 A 59, vol. II, p. 21,4-10 Diels-Kranz, p. 105 Curd.

253. See above, 1184,7-10.

254. Aristotle, *Physics* 8.1, 252a23-4, with curtailments.

255. Aristotle, *Physics* 8.1, 252a26, although Simplicius appears to have read *tauton* instead of the *touto* printed by Ross.

256. Aristotle, *Physics* 8.1, 252a33.

257. Aristotle, *Physics* 8.1, 252a35.

258. cf. Aristotle, *De Interpretatione* 10, 20a20-2.

259. Aristotle, *Physics* 8.2, 252b14.

260. Aristotle, *Physics* 8.2, 252b24-5.

261. Aristotle, *Physics* 8.2, 252b27.

262. Uniform stuffs.

263. Neither Moraux nor Diels-Kranz take up this passage.

264. See Aristotle, *Physics* 3,5.

265. Aristotle, *Physics* 8.2, 252b31.

266. Aristotle, *Physics* 8.2, 252b35.

267. Aristotle, *Physics* 8.2, 252b31-2.

268. Aristotle, *Physics* 8.2, 252b35.

269. Aristotle, *Physics* 8.2, 253a4-5.
270. cf. Simplicius on 8.3, 1193,3-4.
271. Aristotle, *Physics* 8.2, 253a10.
272. Aristotle, *Physics* 8.2, 253a16-17.
273. These *phusikai dunameis* are the result of an elaboration on Aristotelian thought by Galen (for instance, *De Nat. Fac.*, vol. 2, p. 158,5ff.; 177,9ff. Kühn), which was to have a huge influence on medieval and Renaisssance thought. The attractive faculty explains how parts of the body attract nutritive elements from the blood; the retentive capacity explains how substances can remain at a given place in the body; the alterative capacity accounts for how food is converted into blood; and the expulsive capacity explains secretions and expulsions.
274. Aristotle, *Physics* 8,2, 253a13.
275. Aristotle, *Physics* 8,2, 253a15.
276. Aristotle, *On Sleep* 3, 456b21ff.
277. Aristotle, *Physics* 8,2, 253a21.
278. At 253a3-5.
279. *onta*, a plural form of the present participle of the verb to be, can be a little difficult to translate. The most literal rendering, 'beings', is unsatisfactory because it is normally used of animate beings whereas *onta* frequently, as here, also embraces inanimate objects. On the other hand, circumlocutions such as 'things there are', 'things that exist', 'existing things' are cumbersome and often seem to involve over-translation, especially when the word recurs frequently, as here. Actually, it seems to me that in the present chapter *onta* means little more than 'things' (an impression heightened by the fact that Aristotle uses *panta* ('all [things]') even more frequently in the course of the argument), and I have for the most part settled for that rendering. (For anyone who is interested, relevant instances of *onta* occur at 1193,7.12.13.14.18.19.37; 1194,2.5.9.13.15.24.30; 1196,7.18; 1200,30.32; 1203,13.34; 1204,11.21.39; 1205,2.4.39; 1206,1.10.13.25.)
280. 'the more precise articulation of the foregoing' seems to be an interpretative expansion of Aristotle's 'investigation' (1193,1).
281. At 253a4-5.
282. At 253a5-7.
283. Simplicius believes that Aristotle is moving straight into the clarification of the issues surrounding animal motion that he promised us at the end of the previous chapter, but, as modern editors usually indicate, the reference there appears to be to 259b1-16.
284. Posed at 250b11-15.
285. As the sequel confirms, those raised in Chapter 2 (252b7-253a28).
286. Perhaps Simplicius has in mind 253a31-3.
287. Or simply 'from the beginning' or 'from further back'. A more literal rendering would be something like, '... carrying out a division from the top, determine how things stand ...'.
288. Changing both instances of *kineisthai* at 1193,15 to *kineitai*; the indicatives have presumably been assimilated to the infinitives after *anankê*.
289. This summary of Aristotle's division is a mixture of quotation and (often close) paraphrase. As the added numbers and letters are meant to bring out, Simplicius seems to believe (but cf. 1200,29ff.) that the division is into three sections with a further three-way division of the second section, but it seems that it is actually the third section that Aristotle subdivides, yielding the following arrangement at 253a24-30: 'Now it is necessary (1) that all things are either always at rest, or (2) that all things are always in motion, or (3) that some are in motion and some at rest; and again, in the case of these last, either (a) that those that move are always in motion and those that rest [always] at rest, or (b) that all

things are by nature equally inclined to move and to be at rest, or that the only remaining (and third) [option applies]. For it is [also] possible (c) that some things are always motionless, others always in motion, and that [still] others partake of both [these conditions]'.

290. As stated earlier, the reference is to the 'objections' to the position that motion is everlasting in Chapter 2. At 1186,38-1188,16 Simplicius identified three objections there, at 252b9-12, 252b12-16 and 252b17-28 respectively. In the present passage (a) clearly relates to the first of these and (b) to the second, but (c) seems to look to Aristotle's reformulation ('why is it not the case that some things are always at rest and others always in motion?') of the second objection at 253a5-7, which is repeated at the beginning of the present chapter. This in fact makes sense. Although the reformulation occurs in the rebuttal of the second objection, it is equally applicable to the third, which, despite Aristotle's 'third' at 253a8, is really only a special case of the second.

291. Or perhaps, in view of *ek tinos eis ti* at 252b10, 'from one thing to another'.

292. The words 'arises without having existed [previously]' take up a similar phrase at 252b16. When Simplicius similarly paraphrased the same phrase five times between 1187,30 and 1188,5 he actually added 'previously' (*proteron*), so it is a little surprising that he doesn't here.

293. In what follows *telos* seems to hover between 'conclusion' and 'goal' and my 'culmination' at 253a31 in the lemma is something of a compromise.

294. For Simplicius this normally comprised Books 6-8 (cf. the note on 253b8 in the lemma after next), but note that in the parallel passage at 1195,28 he talks of the 'treatise on physics'; the *telos* is, as he spells out at 1193,35-1194,1, the vindication of the final subdivision of the division under discussion, with all that implies. Interestingly, in his introductory remarks to the present book, clearly also with an eye to the present passage, he says (1117,7-8) that Book 8 constitutes 'the best conclusion' to the treatise (*pragmateia*) on the physical principles (which must there mean the *Physics* as a whole), and Graham, seemingly also with this same passage in mind, begins his introduction with the sentence: '*Physics* VIII can be viewed in the way that Aristotle himself may have viewed it, as the crowning achievement of his theory of nature.'

295. A lot to milk out of a present participle but I think it's the correct nuance.

296. For the Pythagorean idea that the number three, the first number that can be said to have a beginning, middle and end, is the number of wholeness and perfection, cf. Aristotle, *Cael.* 268a10-13 with Simplicius comments at *in Cael.* 8,22ff. There may also, as an anonymous reader suggests, be a reference to the progression in the *Physics* from sublunary things, which are sometimes in motion, sometimes at rest, to supralunary things, which are everlastingly in motion, to the unmoved mover at the end of the treatise. The plurals 'beginnings' and 'ends' are in any case puzzling. Elsewhere (and similar phrases are not uncommon in the commentaries; cf. for example 1193,33 a few lines back) Simplicius uses singulars, and it is difficult to see the point of the plurals here, especially perhaps of 'ends'. (*arkhai* of course suggests 'principles', but the principles in the *Physics* relate solely to the sublunar world and shouldn't form part of a progression from the sublunary world to the unmoved mover. Perhaps, however, the plurals are 'corrections' by someone who saw a reference to them.)

297. Deleting *tôi sôritêi* at 1194,12-13. Diels' text would have to mean something like, 'the third is also true by sorites', which, as far as I can see, makes no sense. In view of the fact that the variant *sôtêri* (the dative case of 'Saviour') occurs as a variant in one manuscript, I suspect that the conjunction of 'third' and 'true' inspired a Christian scribe to think of Christ, the third member of the Trinity (for *alêthês* as an epithet of the Son, see Lampe, *alêthês*, B) and to write 'Saviour' in the margin or above the line, and that a later scribe tried to make sense of this by converting it to 'sorites'. (For the idea that the third subdivision is the true one, cf. 1207,17ff.)

298. 'At once' is odd. One would expect 'sometimes in motion and sometimes at rest', as earlier.

299. At 253a29.

300. The terminology and the distinction are of course Aristotle's own; see, for example, *Phys.* 226b10-17. (Note too the importance of the distinction to Simplicius' comment at 1195,37-1196,4.)

301. Translators interpret 'a whole' and 'a part' here (Hardie and Gaye have 'a whole system, not a particular detail', Graham 'the whole of experience rather than some part of it'), but, although the sense at first seems clear enough, it is not actually obvious what the whole in question is. To judge from 1195,1-6, Simplicius seems to believe that it is nature or 'natural things' (*ta phusika).*

302. More literally, 'relating to what was just now said'.

303. At 253a30-2.

304. cf. 1193,32 and the note there.

305. cf. 1193,33-1194,1, where Alexander's description of the *telos* has rather more in common with its description here than does Simplicius' own. The 'lowest' things are the contents of the sublunar world (which are sometimes in motion and sometimes at rest), the 'highest' the contents of the heavens (which are always in motion), and the transcendent principles (which are always at rest) are (or include) the unmoved mover(s); cf. the similar language at 1359,5-8 and 1366,17-22.

306. Or perhaps 'criticises', 'refutes'.

307. From Hellenistic times onwards *kritêrion* was a semi-technical philosophical term for any means or instrument of evaluation, especially of things that can be characterised as true or false. A standard study is G. Striker, '*kritêrion tês alêtheias*', *Nachrichten der Akademie der Wissenschaften in Göttingen, Philologisch-Historische Klasse*, Jahrg. 1974, Nr. 2, pp. 47-110, English tr. in her *Essays on Hellenistic Epistemology and Ethics*, Cambridge 1996, 22-76.

308. Punctuating with a full stop rather than a comma at 1194,28.

309. I suspect that 'motion' (*tên kinêsin*) is a marginal gloss on 'a part' (*meros ti*) that has found its way into the text.

310. Or, to paraphrase, 'if the essence of nature lies it its being ...'. Phrases following this pattern occur a number of times in Simplicius and his *bête noir* Philoponus.

311. Or perhaps something like 'attacking' or 'making difficulties for', but the usage of the verb and the construction are rather against either.

312. Adding *ou* before *tinos* at 1195,10, as the argument would seem to demand.

313. As often in such cases, the expression 'those around Parmenides and Melissus' seems to be little more than a circumlocution for 'Parmenides and Melissus'.

314. Simplicius has earlier argued that the subject of the first part of Parmenides' poem and the writings of Melissus is the intelligible word and that Aristotle understood this, and was in essential agreement with their thrust, but, largely for pedagogical reasons, chose to criticise their arguments as though they were denying the existence of motion in the physical world. The relevant passages, and the origin of these and other similarly implausible positions in Simplicius' desire to produce a 'harmonising' reading of the history of Greek philosophy, are discussed in P. Golitsis, *Les Commentaires de Simplicius et de Jean Philipon à la Physique d'Aristote: Tradition et Innovation*, Berlin 2008, 89-108.

315. At 185a1-3.

316. *dokountas* ('appearing') has its problems. On the one hand, at 185a5-12

Aristotle describes the arguments of Parmenides and Melissus as eristic, which could suggest that he believed that they were not seriously intended, on the other, from Simplicius' perspective, and, he would argue, Aristotle's, they are not only seriously intended but correct, though restricted to the intelligible world. Perhaps he should have added the qualification 'sensible' or 'physical' to 'being'.

317. For this equation in Aristotle, see, for example, *Metaph.* 1013a14-16, *EN* 1151a16-17.

318. sc. the method of the physicist.

319. sc. than maintaining that all things are at rest.

320. 192b20-3.

321. This is one of a number of references, both within and outside the *Physics*, that suggest that it was not originally composed as a single work. Simplicius argues that it in fact consists of five books (1-5) on principles entitled *ta phusika* (the *Physics*) followed by three *peri kinêseôs* (*On Motion*), but the truth seems to be rather more complex. For details, see Ross, *Introduction*, 1-19, who argues for an original *Physics* (Books 1-4), an original *On Motion* (Books 5, 6 and 8), and a 'comparatively isolated' (p. 4) Book 7.

322. As both Simplicius (1197,17.25; 1198,19) and modern translators see, both here and in the next sentence 'divided' doubtless stands for 'divisible'.

323. At 1197,31 Simplicius glosses *phthisis* in this passage as a product word and I have translated accordingly. There is something to be said for this, since Aristotle has indeed been talking about the divisibility of the material removed by a drip, but *phthisis* is normally a process word ('wasting away', 'diminution', 'decrease', etc.) and seems to be such in its other sixty or so occurrences in Aristotle (and, for that matter, at 1196,25 and 1198,28), and modern translators are doubtless right to treat it as a process word here.

324. Although not only human beings can sicken, the neuter is a little surprising and in fact most manuscripts have *tis* ('anyone', 'someone') and that is what most translators translate. However E, the best manuscript of the *Physics*, and Ross have *ti* ('anything', 'something') and, more importantly for present purposes, as 1199,25ff. shows, that is what Simplicius read, so that is what I translate.

325. This is difficult. Perhaps the idea is that the movement from sickness to health cannot be regarded as a single, smooth, continuous, infinitely divisible process, since it involves a transition and that transition is not instantaneous but always involves the intermediate state (cf. b13-14) of convalescence, which can be regarded as a break in the process.

326. The Greek idiom is quite different from the English here and a more literal translation would be something like: 'it would be remarkable if the stone has escaped observation falling or resting on the ground'.

327. The words 'that it is impossible either for all things always to be in motion or for all things always to be at rest' are not in the part of the lemma that Simplicius cites, but the word order makes it difficult to show this by bracketing.

328. 192b21.

329. Punctuating with a comma rather than a full stop after *akinêton* at 1196,4.

330. 1195,30-1196,5 are an intelligent expansion of Aristotle's brief remarks at 253b6-9.

331. It is of course likely that Aristotle had the Heraclitans (or Heraclitus himself) in mind. (Graham has a good discussion of the extent to which they were actually suitable targets for Aristotle.) Democritus, however, is a less plausible target, since none of Aristotle's arguments address the possibility of continuous motion at the atomic level.

332. Not quite a literal quotation; Simplicius inserts the words translated 'all things' (twice) and 'indeed'.

333. More literally, 'nor to join growth to decrease'.

334. Although *stalagmos* is normally, as here, to be translated 'dripping', in what follows it often needs to be translated 'drop'.

335. Here we have the opposite situation. *rhanis* normally means 'drop', but here a case could be made for 'dripping'.

336. Choerilus fr. 11 Bernabé. Choerilus of Samos was an epic poet of the fifth century BC. According to the excerptor of the second half of Philoponus' commentary on the *Physics*, he also quoted this line but attributed it to the second-century BC poet Moschus, and Galen also cites it twice but without attribution. The sentiment was proverbial and *TLG* searches turn up a number of similarly worded versions of it.

337. More literally, 'lessened', 'decreased'.

338. More literally, 'decrease', 'lessening'.

339. The distance between the tips of the thumb and the little finger when the fingers are extended.

340. More literally, 'having come forward'.

341. There were 600 drams to a talent and in the most widely used Greek system of weights, the Attic-Euboic, a talent was equivalent to about 25.86 kg.

342. A rather awkward sentence. I'm tempted to delete the second 'always', which is omitted by one manuscript and the first edition, and to take *khôris*, which actually only occurs once and at the end of the clause, as merely amplifying the idea of 'away in 'taken away' (*aphaireisthai*), but its occurrence a few lines later (line 30) with full value is against it.

343. On this interpretation of *phthisis* at 253b22, see the note on the word in the lemma.

344. One might have expected a more qualified statement, something like: So it is clear that, if one thing is affected by another, it will not *necessarily* be affected by any particular portion of it, and that, if one thing is affected by another during a given [period] of time, it is not *necessarily* also affected during every part of that [period of] time but *perhaps* all at once during some one of them.

345. 'Asked' because, as formulated here, the argument culminates in a question.

346. On the sorites, see for example D. Hyde, 'Sorites paradox', *Stanford Encyclopedia of Philosophy* (Fall 2008 Edition), Edward N. Zalta (ed.), URL = http://plato.stanford.edu/archives/fall2008/entries/sorites-paradox/. It isn't necessarily the case that Aristotle, or even Simplicius, or his source, was actually acquainted with any such sorites, and nor, given that sorites arguments seem to have been invented by Eubulides of Miletus, a contemporary of Aristotle, is it clear who the 'sophists' in question are. Interestingly, Alexander calls the sorites a 'sophism' (*Quaestiones* F1-2, Sharples 1994, 89-94; cited in H. Baltussen, *Philosophy and Exegesis in Simplicius: The Methodology of a Commentator*, London 2008, p. 116), so perhaps this whole discussion of the possible relation of the present argument to a sorites goes back to him.

347. At 1198,1-5, accenting *ara* with a circumflex rather than an acute and removing the colon after *legêi* in line 1, removing the full stop after *katatribein* in line 2 and changing the full stop after *analogon* in line 5 to a question mark.

348. Perhaps *doxêi* is in the dative because it takes up the dative *tôi* at 253b15 on which Alexander is commenting.

349. If Simplicius' reading of Alexander is correct, 'drop' is presumably equivalent to something like 'fall' here.

350. Alexander obviously, and probably correctly, believes that the two Greek words are etymologically connected.

351. Punctuating with a comma rather than a full stop after *akousai* at 1198,12 and with a comma rather than a semicolon after *esti* at 1198,13.

352. The implication is, I think, that Alexander would surely have read it this way if he hadn't been influenced by the ship-hauling parallel.

353. sc. the water.

354. sc. whichever of the two forms the sorites took.

355. Punctuating with a comma after *teinei* at 1198,21.

356. Reading *eite* rather than *ei te* at 1198,23.

357. Notice that although this sentence ultimately depends on 253b21-3, where Simplicius reads *phthisis* as a product word (and although he still sees that passage as being concerned with the divisibility of the product), it is clearly a process word here.

358. 'These things' (*tauta*, 1198,32) are presumably things that grow larger or smaller.

359. 'Something' (*ti*) is only present in one manuscript and is not in Ross's text.

360. More literally, 'prepares the weakness of the rock in advance for the splitting'.

361. I would like to translate 'once they attain maturity' (cf. 1201,39), but that doesn't work for 'decrease'.

362. The Greek of the last clause is unsatisfactory and I have emended *genomenon* (lines 3-4) to *genomenê*. Another possibility, suggested by a reader, would be to add *ti* ('a thing') before *proeisin* (line 3) and translate: 'and although up to a certain point a thing proceeds gradually without being observed, it comes to a complete halt once it has come to consummation upon reaching the goal towards which its growth or decrease has been striving'.

363. Both 'curdling' and 'freezing' translate *pêxis* at 1199,14.

364. A more literal translation would be something like: 'least portion by least portion'.

365. Deleting the comma after *phêsin* at 1199,16.

366. 186a15-16; Diels also compares 966,15; 978,35 and 989,23.

367. But perhaps the quotation only extends to the end of the previous sentence (Diels doesn't close the quotation marks).

368. For this (rather unexpected) 'anything', see the note on *ti* at 253b25 in the lemma.

369. sc. and not from anything else.

370. Simplicius replaces Aristotle's *phanai* with its synonym *legein*.

371. In the lemma I translated '[things] alter', but the singular works better here.

372. *leukansis* ('growing white') and *leukos* ('white') are cognates in Greek, the preposition *ana* ('up') forms part of the compound word *anaphora* ('upwards motion'), and growth (*auxêsis*) and increase (*prosthesis*) are virtual synonymous in some of their applications, but it is not obvious how generation (*genesis*) can be said to be 'named from' being (*ousia*). Perhaps the point is that *genesis* is often described as a movement from non-being to being (cf. 825,6-7, for example), in which case something like 'defined in relation to' would have worked better in its case.

373. 1200,3-9 = fr. 112 Wehrli.

374. More literally, 'the stone certainly has not escaped us at one time falling, at another resting on the ground'.

375. Or 'can move'.

376. I have translated *phtheiretai* ('ceases being') twice with two different subjects, which is what Simplicius' interpretation at 1203,21-4 seems to require. This is awkward (although Hope translates in much the same way), but Aristotle's Greek is difficult anyway. (Hardie and Gaye, Graham, and Cornford (in note ad loc. in Wicksteed and Cornford) treat *phtheiretai* as a copulative verb and translate *phtheiretai touto* 'it ceases to be that', making the phrase the opposite of *gignetai touto* ('it comes to be that') earlier in the sentence, so that the whole clause might be translated 'and ceases being, or being in, what it changes from', which is probably as good as one can do).

377. 'There are things that are sometimes in motion and sometimes at rest' would be clearer.

378. For this, cf. 1193,16.

379. sc. he goes straight to the first subdivision of the section, as he also did in the case of the first section, as Simplicius points out at 1194,30-1.

380. cf. 1193,16-17.

381. Punctuating with a comma rather than a full stop after *kinêsin* at 1200,37.

382. Although, following Diels, I have treated it as paraphrase, what follows (as far as line 19, or even line 25) could be direct quotation.

383. Punctuating with a comma rather than a semi-colon after *phthora* at 1202,1.

384. Adding *hoion* before *leukos* at 1202,19.

385. Simplicius' paraphrase at 1203,9-11 suggests that he read the sentence this way. Modern scholars construe it differently: Graham, for example, translates, 'We must begin again from our recent distinctions, adopting the same starting-point as before'.

386. At this point Ross inserts the words *ê ta men aei êremein ta de aei kineisthai* ('or that some are always at rest and others always in motion') into his text to bring this passage into line with 253a24-30, which it recapitulates. Simplicius observes the omission but assumes that it is deliberate and (at 1203,11-17) suggests possible reasons for it.

387. 253a24-30.

388. 254a3-15.

389. 253a27-8.

390. At 254b1-4.

391. sc. what is fundamental.

392. 253a32-b6.

393. 'Persistence' (*monê*) has been added by Diels on the basis of 1204,35.

394. Another Stoic term.

395. Aristotle has 'but many things seem to be in motion'.

396. There is a certain irony here. Although this is a concession on a straightforward reading of Aristotle's criticism of Melissus, on Simplicius' reading of it (for which see the note at 1195,15) Aristotle should have nothing to concede.

397. For 'those around' see the note at 1195,13.

398. *pseudôs doxazesthai* ('falsely opined') is a deliberate echo of Aristotle's *doxa pseudês* ('false opinion') at 254a27.

399. The verb (*peritrepein*) is sometimes best rendered 'refute' in contexts involving argument, but here the reversal is literal (cf. 'run in the opposite direction' (*anapalin ... hodeuei*) five lines later). Similarly, at 1205,5 the cognate noun *peritropê* is used of the literal reversal of Melissus' argument whereas LSJ only cites passages where it is used of '*turning* an opponents arguments against himself'.

400. A more obvious translation would be 'that according to perception many things appear to be in motion', but the argument seems to turn on the acceptance of the appearance as veridical rather than on the acknowledgement of its occurrence.

401. The followers of Parmenides and Melissus.

402. Changing the second instance of *mê on* at 1205,28 to *kenon*.

403. The words of a hypothetical objector.

404. This may be intended to cover both changes of mind on the part of individuals and the ebb and flow of beliefs in the population at large.

405. More literally 'go round and round'.

406. Or perhaps 'texts'.

407. cf. the comment of the hypothetical objector at 1204,32-3.

408. 'Without [explicit] warrant' is an attempt to make sense of *apistos* at 1205,13, which should by rights mean something like 'not credible', 'without any warrant', but Diels may be right to suspect that it is a (misguided) intrusion into the text.

409. 'Common conceptions', or 'common notions', entered philosophy with the Stoics. They were general ideas held by everyone in areas such as religion, ethics and mathematics and were either innate (unlikely for the Stoics themselves) or formed early and inevitably in interaction with the world. Example of immediate propositions based on such conceptions would be 'god is good' and 'parallel lines never meet'.

410. Changing *pepisteumenon* to *pepistômenon* at 1205,33.

411. 'Things' (and 'them' and 'the others' in what follows) reflects the lack of precision of the Greek. On another day Simplicius might have written 'section' (followed by 'it' and 'the other sections').

412. The second of the two explanations already offered at 1203,11-17.

413. I'm tempted to emend *houtos*, which is a rather surprising way of referring to Aristotle at this point, to *houtôs*, and to translate: 'on the basis of which he also states that [their] similarity is so obvious'.

414. Or 'from the division'.

415. 1206,23-9 = fr. 113 Wehrli.

416. More literally: 'those divided two-ways are also refuted'.

417. There is one minor deviation in Simplicius' indication of the quote from the text of the *Physics* in the manuscripts: Simplicius has *pan*, the *Physics hapan* (both words mean 'the whole', so the change in reading is a matter of diction).

418. cf. *Physics* 3.2, 202a3-5.

419. cf. *Physics* 5.1, 224a21-34.

420. The translation follows the text of the manuscripts as reported in Diels' edition. An alternative is, as Professor Chase suggested, to change the infinitive *sumbainein* to the third person singular *sumbainei*, and translate 'The principle of this motion happens to be somehow in the animal itself, he makes a distinction, saying that the animal moves then by nature, but nothing hinders that the body of the animal ...'.

421. 254b22.

422. 254b23-4, Aristotle's text has instead of the conjunctions *te ... kai* (both ... and) only a single *kai* (and).

423. 254b28-9.

424. cf. 254b29-30.

425. 254b14-15.

426. 255a1-2.

427. The phrases come from 255a2-4.

428. *Laws*, 895C.

429. *Phaedrus,* 245E, Simplicius leaves out the nominal phrase governing the clause – 'every body, for which motion is from outside ...'.

430. Simplicius recasts the argument in the syllogistic form *Camestres*, giving the premises in transposed order, minor premise first, followed by that major premise. 'No *C* is *A* [ensouled]; Every *B* [that which moves itself] is *A* [ensouled]; Therefore: No *C* is *B* [that which moves itself]'. Note, however, that on Simplicius' recasting neither the premises, nor the conclusion contain an explicit quantifier.

With the affirmative major premise the quantifier most probably has to be tacitly assumed – the premise asserts that *everything* that moves itself is ensouled. The case of the negative minor premise, and of the negative conclusion may be construed along the lines of the stock example of *Barbara* (Every man is mortal; Socrates is a man; Therefore: Socrates is mortal), where the minor premise and the conclusion are affirmative sentences with an individual term, and hence they do not contain a quantifier.

431. 255a9-10.

432. 255a10-11, Aristotle's text is slightly different, having a further pronoun (*auta heauta*), stressing the reflexivity of causing self-motion even further.

433. Alexander does not suggest a different reading, rather he submits that the particle *de* should have a causal force here.

434. Apparently Simplicius casts Aristotle's reasoning in the syllogistic form of two conjoined *Barbara* syllogisms, where the premises of the second syllogism ('What are continuous in the strict sense are not affected by themselves', and 'What are not affected by themselves are not moved by themselves') contain negative terms.

435. 255a16.

436. Are always moved – it is always the case that some of them are moved; and it is always the case that they are moved by something whenever they are moved.

437. 255a19-20 – here the word 'moving' is Simplicius' disambiguation.

438. This clumsy clause translates the reading of the manuscripts as reported by Diels (*all' esti biâi, hôs hê toiautê kinêsis*). The alternative is to introduce an emendation, as suggested by Professor Chase, changing *biâ hôs* into *biaios*. With this reading the sentence would mean 'but such movement is constrained'.

439. Simplicius' explication of the phrases 'movers that move naturally' and 'movers that move contrary to nature' sets primary emphasis on the fact that the movement caused by a mover is contrary to the nature *of the moved entity*, and only at the end of this passage returns to consider the further stipulation, that the movement is also contrary to the nature *of the mover itself*. Note that such a scheme leaves unaccounted for the cases where a mover is the cause of some constrained motion *in accordance with its own nature*, but nevertheless the motion is *contrary to the nature of the moved entity*.

440. The text of the manuscripts as reported by Diels apparently needs to be emended. For this translation the infinitive *kinein* (to move) is changed into *kinêtikon* (mover). An alternative would be to change the nominal phrase from the genitive *tou dunamei toioutou* into accusative *to dunamei toiouton*, the translation with that text would run: 'What is thus-and-so in actuality is said to move naturally what is such in potentiality'.

441. 255a25-6. The first part of Simplicius' quote differs from the reading of the manuscripts in the spelling of the reflexive pronoun (*heuatôn* in Simplicius, for the form *hautôn* of the manuscripts of the *Physics*). In the second part Simplicius' text includes the pronoun 'this' (*tautên*): this may be a variant reading, or it may be inserted by Simplicius, due to the fact that he made a remark about the sentence, thereby cutting it into two phrases, which may necessitate the introduction of this anaphoric pronoun.

442. Gardens of Adonis are apparently plants sown in special tubs where they grow and wither at an extraordinary speed.

443. Either Simplicius read 'are changed previously' (in Greek these words are at the end of the clause), or he supplied these words. In the Aristotelian manuscripts the sentence ends with a clause without a verb – i.e. the verb has to be supplied from the previous clause, which would mean that they are moved into their actuality, what they [already] are potentially. A further textual variant: the Simplicius manuscripts read *heautôn* (their own), whereas the manuscripts of the *Physics* have *hautôn*.

444. The manuscripts of Simplicius have *hoper* (that which, what), in singular, as if different changes would be directed at the same single goal. An easy change, suggested by Professor Chase, to read *haper* (those which) in plural could easily mend this problem.

445. Simplicius leaves out the word *heautôn* (their own).

446. Simplicius uses the idiom of an entity receiving something – its motion, its substance – from some other entity to express the fact that this second entity brings the first entity both to being the substance it is, and to the motion which is natural to it.

447. 'Somewhere' (*pou*) is the reading of the Simplicius manuscripts at 1216,30 (see below), the Aristotelian manuscripts have 'to somewhere' (*poi*) instead, with that the translation of the clause is 'The cause is that they are of a nature [to be moved] to somewhere'.

448. 255a30-3.

449. i.e. this intermediate state is related to complete potentiality *as actuality*, whereas it is related to actuality *as potentiality*.

450. 255b4.

451. This is the translation of the reading of the manuscripts, *to kaiein te kai mê kaiein*. It is strictly speaking not impossible – fire burns most of the time, but on occasion it does *not* burn – but the reading remains suspicious. Perhaps one should delete the second option, and speak only about the capacity of fire to burn.

452. This is the translation of the reading of the manuscripts. The Aldine edition prints a conjecture, on the analogy of the example at lines 7-9 above, with that the translation would be: 'For learning belongs to something still incomplete, and it is [itself] incomplete'.

453. This is the translation of the somewhat problematic reading of the manuscripts. The text is problematic, because the introduction of two kinds of potentialities did not rest on the universally valid claim that actuality and potentiality have to be strictly correlated pairs. Hence I am tempted to think that the two transmitted clauses both form the *protasis* of this sentence, where the *apodosis* is missing: 'But since before they are in actuality so, they are first such in potentiality, and potentiality is also twofold, as it has been said, [...]'. Such a complex *protasis* could have been followed e.g. by an *apodosis* which states that similar to the duality of potentialities, in these cases there is a duality of actualities, too.

454. This sentence is also problematic as it stands in the manuscripts. The Aldine apparently tried to mend matters by inserting *aêr* at the end of the clause, that change would give the meaning 'it is somehow air in potentiality'. In view of the context, however, Diels was right to label this addition otiose. From the immediately following sentence (and if my remark about the previous sentence was along the right track, also from the previous one) one would expect – as Professor Chase suggests to me – that the clause should be supplemented by the words *kouphon estin*. With that addition the sentence would mean: 'For when it is water, being heavy in actuality, it is somehow light in potentiality.'

455. Simplicius changes the quote slightly: the obstacle in Simplicius' formulation is 'removed' (*aphairôn*), whereas in Aristotle's text it is 'moved' (*kinêsas*).

456. Read *enekhthênai* instead of the typo in the edition *anekhthênai*.

457. The translation takes Professor Chase's suggestion, that the reading of the manuscripts *metaballontôn* (imperfect participle, 'what changes them') should be corrected into *metabalontôn* (aorist participle, 'what effects the change in them, what changed them').

458. See *Physics* 3.3, esp. 202a21-b5.

459. The translation assumes that the reading of the manuscripts (*metaballein*, infinitive) is changed into *metaballonta*, imperfect participle.

460. See e.g. *Physics* 2.8, 199a17ff.

461. 'Intellect' is supplied by Diels from the Aldine edition.

462. Note that this introductory claim will have to be taken in a way that allows that the something else by which an animal is moved is a constituent of the animal, and hence the animal – contrary to this initial formulation – can be said to be moved by itself. Alternatively, one would need to introduce some emendation which restricts the claim of this clause to the motions of the elements.

463. The elements.

464. 256a1.

465. The syllogism is in *Barbara*, with the complex middle term 'moved by nature or contrary to nature'. Note that this complex term seems to be formulated loosely in the major premise with conjunction instead of disjunction 'moved contrary to nature and according to nature'. One excuse for that would be to take Simplicius' syllogistical rendering as not giving the major premise in this case: instead, in a shorthand way, he formulates two statements – (1) that everything that is moved contrary to nature is moved by something, and (2) that everything that is moved according to nature is moved by something – and these two, taken together are equivalent to the major premise of the syllogism, with a disjunctive complex term.

466. 'The first, therefore the second' – this shorthand is the second part of the standard formulation *modus ponens* in Stoic terminology, where the ordinals stand for propositions: If the first then the second, but the first, therefore the second. Note that the first here stands for the conjunction 'Everything that is moved is moved either by nature or against nature and both the [things that are moved] by nature and the [things that are moved] against nature are moved by something'.

467. 'In a more physical manner' is contrasted to a proof which is in a more logical manner (*logikôterôs*), which remains on a level of generality and argumentations that does not take into consideration the philosophical issues involved. Simplicius refers here to *Physics* 7,1-2.

468. 'Because of' translates the preposition *dia* with the accusative case. (Other possible translations would be 'owing to', 'thanks to', 'on account of', 'in consequence of'.) A thing causes motion 'because of itself' when the motion originates within it (cf. 1222,33-4), while a thing that causes motion 'because of something else' does so as a result of being moved by a third thing. This may be the best place to list a number of other constructions that recur over the next few lemmata and that I've tried to render consistently at the cost of occasional awkwardness. *hupo* with the genitive case, which I render 'by', is used to express agency, whether animate or, with mild personification, inanimate. For the so-called instrumental dative (a plain dative without any preposition) I use 'by means of'; other common renderings would be 'by', or 'with'. (In what follows the same entities are sometimes treated as agents, sometimes as instruments.) *dia* with the genitive case, which I translate 'through', is used to express the means or agent used to achieve something. At times in what follows it seems to be more or less equivalent to the instrumental dative.

469. Counting back from the last thing in the series, the thing which is moved but does not impart motion.

470. This doesn't of course exclude the possibility of a single intermediary.

471. Presumably again counting back from the last thing in the series, since the prime mover is by definition 'such'.

472. This parenthetic comment in fact explains why 'there must be some mover which is not first [moved] by another thing'.

473. In the previous chapter; the conclusion is drawn at 255b31-256a3.

474. I normally translate the passive of *kinein* 'to be moved', but occasionally, as here, the context strongly suggests the rendering 'to move' (in the intransitive sense) or 'to be in motion'. In view of his belief (stated in this very sentence) that anything that moves is moved by something else, it is probable that Aristotle was not always sensitive to the distinction.

475. Translating *an ezêtei* ('he would ... have asked') with F and Diels at 1220; *anazêtei* ('he ... asks'), the reading of the *editio princeps* and of M, is defensible in itself (the question at issue is only 'put off' until 256b13-27), but it looks like an attempt to improve on *anezêtei*, the reading of A, which itself looks like the product of incorrect word division.

476. 1220,29-36 = fr. 115 Wehrli.

477. The earlier, not very systematic, division is to be found at 254b12ff.

478. More literally, 'is among existing things'.

479. Or 'things that [both] cause motion and are moved'.

480. Or perhaps 'by being in motion itself'; but neither version really provides the required contrast with 'moved by something else' and one would actually expect something like 'by being moved by itself'.

481. 1222,14-20 = fr. 116 Wehrli.

482. I've translated on the assumption that Simplicius is correcting Eudemus, but he may in fact be agreeing with him, in which case I would translate: 'And the latter too does seem ...'.

483. For this translation of the instrumental dative see n. 468.

484. 'By itself' would be a more natural rendering, but Aristotle is at pains to put the instrumentality of the mover's 'self' on a footing with that of other instruments it may employ and 'by itself' merely suggests 'without assistance'.

485. This first part of the sentence is difficult and modern scholars are divided as to how to construe it. My own preference would be to take *to hôi kinei* ('that by means of which [a mover] causes motion') as the object of *kinein* ('to cause motion'), as do Ross, in his Analysis, and Graham, and translate: 'But it is impossible [for a mover] to cause motion without there being [a mover] which moves that by means of which it [sc. the second of the two movers mentioned] causes motion by means of itself'. However, Simplicius' paraphrase at 1223,15-16 suggests that, like Hardie and Gaye, Hope (I think), and Wickstead and Cornford, he construed *to hôi kinei* as the subject of *kinein*, and I have translated accordingly.

486. Treating this second part of the sentence ('but while ... to infinity') as epexegetic of the first.

487. Or perhaps 'he' (sc. Aristotle).

488. The phrase *mêpote de* ('But perhaps'), which introduces many of the passages in which Simplicius takes issue with Alexander (another seven occurrences in 8.5 alone), is more likely to be motivated by respect and consequent politeness, or even, as Baltussen suggests (*Philosophy and Exegesis in Simplicius*, London 2008, 129; the whole chapter is a useful discussion of the extent and nature of Simplicius' use of Alexander), serve as cover for 'subtle subversion', than be a sign of any diffidence or doubt as to the correctness of his own position on Alexander's part.

489. sc. the earlier one.

490. 256a4-5. For grammatical reasons Ross prints *auto* rather than *heauto* or *hauto* (the reading of the manuscripts as well as of Simplicius) in both lines, but the sense remains much the same.

491. I would place the words *kai gar kai … di' heauto* at 1222,29-31 between dashes (as in the translation) or enclose them within round brackets, or at a minimum replace the comma after *di' heauto* with a semi-colon.

492. Unattributed quotations are of passages from the current lemma.

493. 'Follow back' is unusual for *anagein*, but it, or something similar, seems to be the required sense.

494. Both symmetry and what follows, even in this sentence, would suggest that Simplicius should have written 'that which causes motion by means of something else' rather than 'that by means of which the mover causes motion', but his reading of 256a25-6 in the lemma (on which see n. 485) commits him to the latter formulation.

495. sc. when the mover is an externally moved mover that doesn't initiate the movement 'by means of itself'; the phrase seems to be inspired by *aei heteron* ('one thing after another') in the passage from the lemma quoted in the next line.

496. Removing the comma after *kinêi* in 1223,4 and adding one after *heteron* in 1223,5, with Ross in his text of the *Physics*; 1224,1-3 shows that Simplicius himself read the passage this way.

497. 1223,5-9 = fr. 117 Wehrli.

498. The opposition between *diakrinein* ('distinguish', 'separate'), and *sunaptein* ('bring together', 'combine') seems to show that Simplicius believed that Eudemus' words imply a different understanding of the arguments from his own and this is perhaps best explained if he understood this *kai* as epexegetic, so that 'because of itself or because of another thing' and 'by means of itself or by means of something else' are, at least broadly, synonymous.

499. The phrasing is a little unexpected, since moving something with a stick is a more obvious case of moving it 'by means of something' than moving it with one's hands.

500. sc. not itself.

501. 242b59-243a31 (Book 7, version 1); 242b24-243a1 (Book 7, version 2).

502. This is difficult, but I think that 'prior to that' must mean prior to the thing that uses it as an instrument (*to kinoun*, or 'the thing that causes motion' in the previous line) and that the point is that in such a series as 'man, hand, lever, stone' the lever's need of the man is 'prior' to its need of the hand.

503. sc. by means of something.

504. The point is that at 256a22-3 *hautôi* ('by means of itself') and *allôi* ('by means of something else') were both cases of *tini* ('by means of something'), but here (and earlier at a28) *tini* is contrasted with *hautôi* in a way that makes it equivalent to *allôi*.

505. Actually, Aristotle doesn't use the word 'absurd' in this connection, although Simplicius himself does at 1223,24.

506. Changing *periousan* at 1224,9 to *periiousan* (cf. 1180,14).

507. In *GC* Aristotle actually argues that each of the elements can change into any of the others (see, for example, 331a12-23), though some transformations are more difficult than others. The nearest thing to the cycle in the text that I can find in *GC* is at 337a1-6.

508. sc. by initiating chains of movement from within themselves.

509. sc. according to a regular sequence, for which cf. lines 9-11 above.

510. As it is at 1232,17-32.

511. Because the English word order differs from the Greek, the position of this bracket is only approximate.

512. In 8.1.
513. Themistius' paraphrase may imply a similar, though more complex, relocation of 256b3ff. (cf. the notes at Themistius, *in Phys.* 220,29 and 223,12).
514. 257a27-9.
515. 258a3-5.
516. Actually, it is hard to find plausible antecedents for *toutôn* and *ekeinôn* (which I have translated 'those [arguments]' and 'these'): perhaps it's as vague as 'things here' and 'things there'.
517. At 256b7-8.
518. Punctuating with a comma rather than a full stop after *prôtou* at 1225,4.
519. If correct, equivalent, I think, to 'but of those that are not only moved but move themselves', but one would expect 'as well as moving other things also move themselves'.
520. The phrase 'the destructive *topos* from consequence' is drawn from a terminology that appears to have been developed by Alexander in his *Topics* commentary in the course of an attempt to elucidate certain of Aristotle's *topoi* ('places' in which different types of argumentative strategy are grouped, possibly for mnemonic purposes) in terms of the five Stoic 'indemonstrables', or hypothetical syllogisms, as he preferred to call them. In this terminology the term 'from consequence' (*ex akolouthias*) indicates that the argument is a hypothetical syllogism whose hypothetical premiss takes the form 'if *p* then *q*', and which is therefore of the first or second of the five types, and the fact that it belongs to the destructive (alternatively 'refuting' or 'denying') *topos* tells us that it is in fact of the second type, which later came to be referred to as *modus tollens*. (The first type, the *modus ponens*, is said to belong to the 'constructive' (alternatively 'establishing', 'confirming') *topos*; the phases are actually Aristotle's own, appearing at *Top.* 119a33-4; 152b38-153a2.) The 'division' (*diairesis*) referred to is the disjunction 'either accidentally or *per se*' at 1225,13-14 (Alexander's term for a disjunction in *in Top.* is the related word *to diairetikon*). The argument itself, then, may be symbolised 'If *p* then either *q* or *r*, but neither *q* nor *r*, therefore not *p*'. Given that the terminology used to describe the argument is clearly Alexander's, it is possible that Simplicius is here, as often, drawing on Alexander's *Physics* commentary, and in fact an argument described in much the same terms at 529,34ff. above is attributed to Alexander – although the argument in a passage that parallels this one even more closely (*in Cael.* 227,32-228,2) is not. (The key passage in Alexander's *in Top.* is 165,6-167,2. S. Bobzien (forthcoming), *Alexander of Aphrodisias on Aristotle's Theory of the Stoic Indemonstrables*, is an account of the key role of Alexander in developing the amalgam of early Peripatetic and Stoic logic that, with relatively minor variations, became the norm in late antiquity. Her discussions of the logical terminology of Alexander's commentaries in her sections 1.2 and 2 were invaluable in the preparation of this note. Chapter 2 ('The Greek commentators on Aristotle') of A. Speca, *Hypothetical Syllogistic and Stoic Logic*, Leiden 2001, was also useful.)
521. The demonstration has already commenced and is arguably complete by 257a31.
522. Reading *kinei to* and *katho* with the manuscript tradition rather than *kinoito* and *kath' hauto* with the first printed edition and Diels at 1225,24.
523. Changing *tôi* to *mê* at 1225,28, with the first printed edition (which also makes other changes to this line).
524. Reading *to te* rather than *tote* at 1225,29, on the assumption that *tote* has been read under the influence of *hote* (for which it often paves the way) and that *hote* actually looks back to the words 'a falsehood can follow'.
525. *An. Pr.* 32a18-20.

526. *tou mê kineisthai kinoun* ('that what causes motion is not moved [itself]') at 1225,33 is rather awkward and I'm tempted to emend *kinoun* to *kinountos*, or to add *to* before *kinoun*. (Neither change would affect the translation.)

527. Punctuating with a full stop after *kinêsis* at 1226,1.

528. More literally, 'being at the same time and together [with it]'.

529. sc. in such a way that it is not merely an instrument.

530. This rendering is explained at 1227,27.

531. The phrase, which literally means something like 'through the stick in the middle', prepares the way for the reference to 'the things between the first movers and the last things moved' in the next clause.

532. 243a32-244b2 (Book 7, version 1); 243a2-244a24 (Book 7, version 2).

533. 244b3-245b2 (Book 7, version 1); 244a25-245b18 (Book 7, version 2).

534. Perhaps one should add *tou* before *kinountos* at 1227,14.

535. Changing *kinoun ti* at 1227,14-15 to *kinounti*.

536. The Greek contains a mild anacoluthon which I have preserved in the translation.

537. *oinomeli* was actually, as the name implies, a mixture of honey and wine rather than a true mead.

538. Modern scholars (e.g. Hardie and Gaye, Graham, Wickstead and Cornford, Ross in his 'Analysis') commonly, and correctly, translate *eulogon, hina mê anankaion eipômen* 'it is reasonable, not to say necessary', or the like, but Simplicius seems to understand it in some such way as I have rendered it.

539. sc. the argument was hypothetical rather than demonstrative. (On the meaning of the phrase 'from a consequence', see n. 520.)

540. In this case, effectively an argument from authority.

541. Only the words 'if' and 'not accidentally' are actually present in Diels' text.

542. 'Kind' (Hardie and Gaye, Ross in his 'Analysis' and Hope) and 'form' (Graham) are both reasonable for *eidos* here and later in the chapter, but the less specific 'kind' strikes me as better. In the commentary, on the other hand, no single rendering seems entirely satisfactory. There are passages (1228,37-1229,5; 1229,16-25) that show that Simplicius believes that Aristotle is thinking of species and a case could be made for so rendering *eidos* throughout. However, there are many passages where that would seem artificial and awkward and I've decided to restrict 'species' to passages where it is clearly called for. Elsewhere, I have normally preferred 'kind' but have used 'form' in a few passages where it seems to work better. (Where it seems useful, I alert the reader to a switch from one rendering to another by inserting '(*eidos*)' in the text.)

543. 'This' is more likely, I think, to cover both alternatives, as Ross ('Analysis', p. 438) assumes, rather than just the second, as Graham's translation suggests.

544. Simplicius (1228,37-1229,5; 1235,1-3) believes that these *atoma* are the *infimae species* of motion (teaching geometry, say, or even teaching a particular geometrical theorem, as opposed to teaching in general), but it seems more likely that the point is that one should consider the implications of the theory for actual instances of a type of motion rather than operate at a higher level of generality.

545. The pronouns in this sentence (and in Simplicius' comments) are actually neuter ('something', 'it'), but, like most modern translators, I shall use the masculine forms.

546. This seems to look back to 1225,14ff. rather than to the refutation of the hypothesis under discussion.

547. At this stage he merely shows that it is reasonable, but later he will *prove* that it is so.

548. At 257a33-258b9.

549. Perhaps the second *kinoumenon* at 1228,8 should be emended to *kineitai*.

550. Simplicius has *ei de mê* for Aristotle's *alla mên ei mê* at 1228,10 and *hôs ei mê* for Aristotle's *ei de mê* at 1228,11.

551. '*Per se*' seems to be short for 'a mover that is moved *per se*', and, similarly, 'the accidental' in the next sentence for 'an accidentally moved mover'.

552. Although I've adjusted the translations a little to suit the context, this and the next quotation are direct quotations from the lemma.

553. To judge from Simplicius' comments at 1230,5-19, with the remark 'But this is obviously impossible' at 257a12.

554. One would expect *to didaskon* rather than *didaskôn* at 1228,33, but the translation would remain much the same.

555. On the face of it there is a problem here, since when a thing is being thrown its potential *qua* 'throwable' is actualised and it is no longer potential, but Aristotle himself writes in a similar fashion at 257b6 and Simplicius is doubtless influenced by that passage. He does, however, see that an explanation is required and, commenting on that passage, writes: '[We] say that what is movable is in motion, even when it is still in potency and progressing toward complete entelechy, since motion, although it is the entelechy of what is movable, is an imperfect entelechy, because the object in motion is still in potency, whereas what causes motion is already in act' (1235,14-17 tr. Chase). (Graham, following S. Waterlow, *Nature, Change, and Agency in Aristotle's* Physics: *A Philosophical Study*, Oxford 1982, 244, resolves the apparent problem at 257b6 rather differently.)

556. Or 'being acted on'.

557. Or 'capable of throwing' for *rhiptikos* (with LSJ, which cites only this passage), and 'able to be thrown' (which should be possible for *rhiptos*), but that doesn't seem to be quite what is needed.

558. Not quite a direct quotation since the syntax has been adjusted to the context.

559. It isn't immediately clear what the genus and its species are. Perhaps something like 'things that cause motion and are moved with the same kind of motion', and 'simultaneous teachers and learners of geometry', 'things that throw and are thrown with the same kind of throw', and the like.

560. Presumably, although it isn't immediately obvious, 'teaching geometry' and 'throwing'.

561. Simplicius has *ei de* ('and if'), rather than *ê* ('or'), the reading (with one minor exception) of the manuscript tradition of Aristotle.

562. sc. each kind of motion.

563. This looks back to *ê kath' heteron* ('or with another [kind]') at 256b31 and might mean little more than 'of a different kind', but the nature of the examples of motions that follow (carrying, increasing, and altering, which are blanket terms for motion in the categories place, quantity and quality, the three categories that admit motion in the strict sense; in the *Physics*, compare, for instance, 201a11-15, 223a29-32, 228b19-21, 243a35-9), suggests that *genos* is here, as often, equivalent to 'category' and we are to think of the kinds of motion as belonging to different categories. This isn't really what one would expect and perhaps the examples were merely chosen to provide a more or less plausible example of a closed loop containing only a small number of motions, albeit at a high level of generality.

564. sc. to connect the last term to the first, creating a closed loop.

565. 'The one' and 'the other' are the teacher and learner implied in the first part of the sentence.

566. At 257a14-25.

567. At 225b5-9.

568. At 204a35-206a8.

569. Arguably it would have been better to retain the translation ‘species’ till the end of the paragraph.

570. sc. with one of the motions.

571. sc. if that is the motion that it reverts to.

572. More precisely, of their being moved – and thereby producing motion.

573. ‘The latter’ are the immediate movers, ‘the former’ the earlier movers, and especially the prime mover.

574. I think ‘them’ has to refer to the people who hold the first theory, which is a little odd: one would expect something like ‘by way of their theory’.

575. Punctuating with a full stop rather than a comma after *dedeiktai* at 1230,10.

576. Punctuating with a full stop rather than a comma after *dunamei* at 1230,13.

577. sc. between the ultimate source of the heat and the last thing heated.

578. sc. the ‘things in between’.

579. The prime heaters, so to speak.

580. Changing *kinoumena* to *kinounta* at 1230,27; cf. *kinoumena kinei* and *tôi thermainesthai thermainei* in lines 24 and 25.

581. sc. one or more.

582. Omitting *ê* before *heauto* in the lemma at 1230,31 with the text of Aristotle. (I can’t make good sense of the transmitted text.)

583. sc. acting and being acted on.

584. Although the sentence contains a direct quotation from the lemma, I’ve had to render it rather differently here.

585. I’ve translated ‘according to one form’ rather than ‘with one kind [of motion]’ because later in the sentence the phrase is taken up by the noun ‘health’, but see the next note.

586. One would expect a process word, say *hugiansis* (‘healing’), here, rather than a ‘state’ word like ‘health’ (cf. ‘alteration’ later in the sentence).

587. sc. the outcome.

588. At 1229,16-25.

589. One would in fact expect *kata to auto eidos* rather than just *kata to eidos* at 1231,19.

590. cf. 1231,4-6.

591. We would expect the ‘first thing moved’ to be the one immediately after the first mover, as it is at *Physics* 257a27 and at 1365,4, rather than the last in the series as here, but cf. 1229,35, where *to prôton pheromenon* (‘the first thing carried’) refers to the last thing in the series – and is actually one of the things referred to as such a few lines later at 1230,1.

592. Removing the brackets around *adunaton … sundramein* and punctuating with a colon after *eipôn* and a comma rather than a full stop after *sundramein* at 1231,28-9.

593. Simplicius omits Aristotle’s *aei* (‘always’) at 257a25.

594. sc. by itself.

595. cf. what follows with the similar passage at 1224,6-23, where this one is foreshadowed.

596. In this and the next paragraph, ‘through the agency of’ translates the preposition *hupo*, which I have hitherto translated ‘by’.

597. The same alternatives figure at 1151,8-11.23-5 above in Simplicius’ excursus on Philoponus.

598. On the cyclical transformation of the elements, see the note at 1224,11; for the cyclical nature of generation, see *GC* 2,10.

599. ‘Also’ presumably means ‘like the things it causes, which in turn cause other things’.

600. I've translated Ross's text and rendered it along the lines implied by his Analysis (and by Simplicius' comments), but one could also (1) translate, 'For what is *per se* is always a prior cause to what is itself also *per aliud* (cf. Hope's 'For a factor which is independent ontologically is always prior in the order of explanation to a factor having a dependent status') or (2) assume, as Graham does, that Simplicius read *aition aei proteron* with Ë in line a30 and translate, 'For what is a cause *per se* is always prior to what is itself also [one] *per aliud*'.

601. The lemma, presumably erroneously, ends here.

602. Supplying *mallon* after *autokinêton* at line 37, as Diels suggests in his apparatus.

603. Aristotle, *Physics* 5.4, 228a20ff.

604. Eudemos of Rhodes, fr. 118 Wehrli.

605. Aristotle, *Physics* 8.5, 257b29-30.

606. Aristotle, *Physics* 8.5. 257b3-6.

607. Fr. 119 Wehrli.

608. Throughout this section, the Greek word *eidos* can mean either 'form' or 'species'.

609. Aristotle, *Physics* 8.5, 257a1.

610. Aristotle, *Physics* 8.5, 257b10.

611. Aristotle, *Physics* 8.5, 257b11-12.

612. Theophrastus, fr. 176, vol. I, 336-8 Fortenbaugh et al.

613. Aristotle, *On Generation and Corruption*, 1.7, 324a9-13.

614. Reading *<to kath'> hauto* instead of *auto* at line 36, following a suggestion by Richard Sorabji.

615. Fr. 120 Wehrli.

616. *antanairesis*. Alexander of Aphrodisias (544,17 Wallies) informs us that by this term Aristotle referred to *anthupairesis*, or continuous subtraction, brilliantly studied by D.H. Fowler *The Mathematics of Plato's Academy a New Reconstruction*, Oxford 1987, 2000[2].

617. Aristotle, *Physics* 8.5, 256a10.

618. At line 34, I omit the *prôtên* supplied by Diels from A[1].

619. Aristotle, *Physics* 8.5, 257b16-17.

620. Aristotle, *Physics* 8.5, 257b19-20.

621. Aristotle, *Physics* 8.5, 257b22.

622. Aristotle, *Physics* 8.5, 257b20-1.

623. Aristotle, *Physics* 8.5, 257b25.

624. The words 'neither a single part nor each of a number of parts' should really be inside the curly bracket but the different word-order of the English doesn't allow it.

625. Aristotle, *Physics* 8.5, 258a1-2.

626. Alexander.

627. Aristotle, *Physics* 8.5, 258a12.

628. Aristotle, *Physics* 8.5, 258a7-8.

629. Aristotle, *Physics* 8.5, 258a20-1.

630. cf. Aristotle, *Physics*, 5.3, 227a10; *On Generation and Corruption* 1.6, 323a3ff.; Simplicius, *in Cat.* 123,29ff. Kalbfleisch.

631. Aristotle, *On Generation and Corruption* 1.6, 323a31-3.

632. Aristotle, *Physics* 8.5, 258a19-21.

633. At *Physics* 8.5, 258a21-2, the text read by Alexander, Simplicius and most of the best manuscripts contains a *protasis* (*ei men oun* ...), but no apparent *apodosis*. Ross solved the problem by inserting at the end of this passage, with MS I, the *apodosis hapsetai hekateron hekaterou*, 'they will touch one another'.

634. Aristotle.

635. Aristotle, *Physics* 8.5, 258a27-9.

636. Aristotle, *Physics* 8.5, 258a26-7.

637. Aristotle, *Physics* 8.5, 258a28-30.

638. Aristotle, *Physics* 8.5, 258a30-2 (with omissions and additions).

639. Aristotle, *Physics* 7.1, 241b29ff.

640. The text appears to be corrupt. I conjecture something along the lines of *autos enistamenos tou apo tou merous <tou autokinountos to autokinêton kineisthai, kai to holon ktl>*.

641. Aristotle, *Physics* 7.1.

642. Aristotle.

643. Counting back down towards the prime mover.

644. But notice that Simplicius has *to prôton kinoun* at 1247,24, which is closer to Ross's text and suggests a different translation.

645. Kalbfleisch inserts *akinêton* after *to kinoun* at line 26, but I omit it, following MSS A, F, and M. I take it the meaning is that just as the self-moved is ontologically prior to all things in motion, so within the self-moved the part that causes motion is also prior.

646. Aristotle, *Physics* 8.5, 257a29.

647. Plato, *Phaedrus* 245C.

648. Plato, *Laws* X, 895B, tr. A.E. Taylor, modified.

649. Plato, *Laws* X, 896A, tr. A.E. Taylor. The text as it stands omits lines 896A1-2. Kalbfleisch restored the omitted lines, which he thought had dropped out by *homoioteleuton*, but I have chosen to follow the manuscripts.

650. Plato, *Phaedrus* 245E.

651. Plato, *Timaeus* 37A.

652. Plato, *Laws* X, 896Eff.

653. Aristotle, *On the Soul* 1.3, 406a10.

654. *to ekstatikon tês kinêseôs*. A reference to *Physics* 4.13, 222b16: *metabolê de pasa phusei ekstatikon*, a difficult phrase omitted, for instance, in Hussey's translation of *Physics* 3-4 (Oxford 1983). Hardie and Gaye translate 'it is in the nature of all change *to alter things from their former condition*'.

655. Aristotle, *Physics* 8.5, 257b2-6.

656. Aristotle, *Posterior Analytics* 1, 71a1-2.

657. *autheueretos*, apparently a hapax.

Bibliography

Baltes, Matthias, *Die Weltentstehung des platonischen Timaios nach den antiken Interpreten, Teil I* (= Philosophia antiqua 30), Leiden 1976. *Teil II, Proklos* (= Philosophia antiqua 35), Leiden 1979.

Baltussen, H., *Philosophy and Exegesis in Simplicius*, London 2008.

Bernabé, A., *Poetarum epicorum Graecorum testimonia et fragmenta*, vol. 1, Leipzig 1987.

Bobzien, S., *Alexander of Aphrodisias on Aristotle's Theory of the Stoic Indemonstrables*, forthcoming Oxford 2012; cited from Academia.edu.

Bollack, J., *Empédocle*, 3 vols, Paris 1965.

Burnyeat, M.F., 'Gods and heaps', in M. Schofield and M. Nussbaum (eds), *Language and Logos: Studies in Ancient Greek Philosophy Presented to G.E.L. Owen*, Cambridge 1982, 315-38.

Golitsis, P., *Les commentaires de Simplicius et de Jean Philopon à la* Physique *d'Aristote*, Berlin/New York 2008.

Graham, D.W., *Aristotle*, Physics, *Book VIII*, translated with a commentary, Oxford 1999.

Hardie, R.P. and Gaye, R.K. (trs), 'Physics', in J. Barnes (ed.), *The Complete Works of Aristotle. The Revised Oxford Translation*, vol. 1, Princeton, 1984.

Hasnawi, Ahmad, 'Alexandre d'Aphrodise vs. Jean Philopon: notes sur quelques traités d'Alexandre "perdus" en grec, conservés en arabe', *Arabic Sciences and Philosophy* 4 (1994), 53-109.

Hoffmann, Philippe, 'Sur quelques aspects de la polémique de Simplicius contre Jean Philopon: de l'invective à la réaffirmation de la transcendance du ciel', in I. Hadot (ed.), *Simplicius: sa vie, son oeuvre, sa survie,* Actes du Colloque international de Paris (28 sept.-1 oct. 1985), organisé par le Centre de Recherches sur les Oeuvres et la Pensée de Simplicius (RCP 739 – CNRS) (= Peripatoi Band 15), Berlin/New York 1987, 183-221.

Hoffmeier, James K., 'Some thoughts on Genesis 1 & 2 and Egyptian cosmology', *Journal of the Ancient Near Eastern Society* 15 (1983), 39-49.

Hope, R. (tr.), *Aristotle's* Physics, Lincoln 1961.

Hyde, D., 'Sorites paradox', *Stanford Encyclopedia of Philosophy* (Fall 2008 Edition), Edward N. Zalta (ed.), URL = http://plato.stanford.edu/archives/fall2008/entries/sorites-paradox/.

Inwood, Brad, *The Poem of Empedocles: A Text and Translation with an Introduction*, Toronto 1992.

Lampe, G.W.H. (ed.), *A Patristic Greek Lexicon*, Oxford 1961.

Liddell, H.G. and Scott, R., *A Greek-English Lexicon*, rev. H. Jones; with a New Supplement, Oxford 1996.

Judson, Lindsay, 'God or nature? Philoponus on generability and perishability', in R. Sorabji (ed.), *Philoponus and the Rejection of Aristotelian Science*, London/Ithaca 1987, 179-96.

Kirk, G.S., Raven, J.E. and Schofield, M., *The Presocratic Philosophers*, Cambridge 1983[2].
Laks, André, *Diogène d'Apollonie. Édition, traduction et commentaire des fragments et témoignages. Deuxième édition revue et augmentée*, Sankt Augustin 2008.
Marcovich, M., *Heraclitus: Greek Text with a Short Commentary. Editio maior*, Merida, Venezuela 1967.
Maróth, Miklós, *Die Araber und die antike Wissenschaftstheorie*, Leiden etc. 1994.
Moraux, Paul, *Der Aristotelismus bei den Griechen. 3, Alexander von Aphrodisias*, ed. J. Wiesner, Berlin/New York 2001.
Puig Montada, J., 'Zur Bewegungsdefinition im VIII. Buch der Physik', in G. Endress and J.A. Aertsen (eds), *Averroes and the Aristotelian Tradition*, Leiden etc. 1999, 145-59.
Rashed, Marwan, 'Al-Fârâbî's lost treatise *On changing beings* and the possibility of a demonstration of the eternity of the world', *Arabic Sciences and Philosophy* 18 (2008), 19-58.
Ross, W.D., *Aristotle's* Physics, a revised text with introduction and commentary, Oxford 1936.
Sharples, Robert, *Theophrastus of Eresus: Sources for his Life, Writings, Thought and Infuence. Commentary Volume 3.1: Sources on Physics*, Leiden etc. 1998.
Sorabji, Richard, *Time, Creation, and the Continuum*, London/Ithaca 1983.
Speca, A., *Hypothetical Syllogistic and Stoic Logic*, Leiden 2001.
Striker, G, '*kritêrion tês alêtheias*', *Nachrichten der Akademie der Wissenschaften in Göttingen, Philologisch-Historische Klasse*, Jahrg. 1974, Nr. 2, pp. 47-110; Eng. tr. in her *Essays on Hellenistic Epistemology and Ethics*, Cambridge 1996, 22-76.
Taylor, C.C.W., *The Atomists: Leucippus and Democritus, Fragments. A Text and Translation with a Commentary*, Toronto 1999.
Thesaurus Linguae Graecae [CD-ROM].
Todd, R.B. (ed. & tr.), *Themistius: On Aristotle Physics 5-8,* London/Ithaca 2008.
Usener, H., *Epicurea*, Bonn 1887.
Waterlow, S., *Nature, Change, and Agency in Aristotle's* Physics: *A Philosophical Study*, Oxford 1982.
Wehrli, F., *Die Schule des Aristoteles*, vol. 8, *Eudemus von Rhodos*, 2nd ed,, Basel/ Stuttgart 1969.
Wicksteed, P.H and Cornford, F.M., *Aristotle,* The Physics*, with an English Translation*, vol. 2, London 1968.
Wildberg, C. (ed. & tr.), *Philoponus: Against Aristotle on the Eternity of the World,* London/Ithaca 1987.
Zahlfleisch, Joh., 'Einige Corollarien des Simplicius in seinem Commentar zu Aristoteles' Physik (ed. Diels)', *Archiv für Geschichte der Philosophie* 15 n.F. 8 (1902), 186-213.

English-Greek Glossary

abasement: *huphesis*
able, be; can, may be: *dunasthai*
absurd: *atopos*
absurdity: *atopon*
accepted: *homologoumenos*
accomplish: *anuein*
accord with: *harmottein*
accordance with, in: *sumphônôs*
account (n.): *logos*
acquire: *apolambanein*
acquire together: *sunapolabein*
act (n.): *energeia*
action: *drasis*
active: *energêtikos, poiêtikos*
active, be: *energein, poiein*
  be in activity: *energein*
  exert activity: *poiein*
activity: *to energein, energeia*
actuality: *energeia, to energeiâi, entelekheia*
actual/in actuality, be: *energein, energeiâi,*
  in actuality, actually: *energeiâi, kat' energeian*
  according to actuality: *kat' energeian*
add: *epagein, prostithenai*
added, be: *proskeisthai*
addition: *prosthêkê*
adduce: *epagein, paratithenai, propherein*
adjacent, be: *geitniazein*
administer: *dioikein*
admit: *epidekhesthai, katadekhesthai*
advance (v.): *proerkhesthai, prokoptein*
affect (n.): *pathos*
affected, be: *paskhein*
  that which can be affected: *pathêtikos, to pathêtikon*
  not affected, cannot be affected: *apathês*
affection: *pathos*
agree: *sumpheresthai, sumphônein*
agreed, be: *keisthai*
agreement, in: *sumphônos*
all: *pas*
all at once: *athroos*
allotted, be: *lankhanein*
alteration: *alloiôsis*
alterative: *alloiôtikos*
always: *aei: pantôs, pas*
animal: *zôion*
apparent: *phaneros*
appear: *dokein, phainesthai*
apperception: *sunaisthêsis*
appetite: *orexis*
apply: *epiballein, prosagein*
apprehension: *antilêpsis, katalêpsis*
approach: *methodos*
appropriate to, be: *prosêkein*
apt: *epitêdeios*
aptitude: *epitêdeiotês*
argue against: *antilegein*
argue dialectically: *epikheirein*
argument *epikheirêma, logos*
arouse: *egeirein*
articulate: *diarthrôsis*
articulate (v.): *diarthroun*
arrive: *katantan, pareinai*
arrogance: *megalaukhia*
ask: *paraitein*
asleep, be: *koimasthai*
assert: *eipein, legein, phanai*
assume: *lambanein*
assuredly: *pantôs*
attach: *exaptein, prosaptein*
attempt (n.): *epikheirêma*
attempt (v.): *epikheirein*
attractive: *heltikos*
avoid: *ekklinein*
avoidance: *ekklisis*
aware of, be: *eidenai*
axiom: *axiôma*

baffle: *sphallein*
ball: *sphaira*
become: *ginesthai*
begin: *arkhesthai*
beginning: *arkhê*
beginningless: *anarkhos*
begininglessness: *anarkhon*

being about: *apergazesthai*
between: *metaxu*
boast (v.): *neanieuesthai*
bodily: *sômatikos*
body: *sôma*
book (n.): *biblion*
bounce off: *apopallesthai*
brain: *enkephalos*
bring: *paragein*
bring together: *sunairein, sunaptein*
burn (v.): *kaiein*

call (v.): *kalein, legein*
capable: *dunatos*
capacity: *dunamis*
carried, be: *pheresthai*
  carried up, be: *anapheresthai*
category: *katêgoria*
cause (n.): *aitia, aition*
cause rest: *êremizein*
cause, specify as the: *aitiasthai*
cease: *pauesthai*
censure (v.): *euthunein*
certainly: *pantôs*
change (n.): *metabolê*
change (v.): *metaballein, methistanai*
change, cross-wise: *parallaxis*
change, qualitative: *alloiôsis*
chatter (n.): *adoleskhia*
choose: *proairein*
circle, come/go round in a: *anakukleisthai*
cite: *paratithenai*
circle: *kuklos*
circular motion: *kuklophoria*
  body which moves in a circle: *kuklophorêtikos*
clear (adj.): *dêlos, phaneros, saphês*
  not clear: *adêlos*
clear, be: *phaneroun*
clear, become: *dêloun*
clear, make: *dêloun*
clearly: *dêlonoti, saphôs*
climb (v.): *anarrhikhasthai*
co-exist: *sunuparkhein*
cognizance, take: *gnorizein*
coincide: *suntrekhein*
coincident: *sundromos*
coldness: *psukhrotês*
combination: *sunkrisis*
combine: *sunagein, sunkrinein*
come about: *ginesthai*
come after: *akolouthein*
come round in a circle: *anakukleisthai*
come to be, come into being: *enginesthai, ginesthai*
come to rest: *histanai*
coming into being: *genesis*
comment (v.): *hupomimnêskein*
commentary: *hupomnêma*
common: *koinos*
common feature: *koinotês*
compacted, be: *sumpileisthai*
complete (adj.): *pantelês, teleios*
completely: *teleôs*
completeness: *teleiotês*
complete a revolution: *apokathistasthai*
compounded, be: *sumpêgnunai*
compressed, be: *puknousthai*
conceive: *noein*
concept: *ennoia*
concisely: *sunêirêmenôs*
conclusion: *sumperasma*
condition: *hexis*
confirm: *pistoun, pisteuein*
confirmatory: *bebaiôtikos*
confound: *sumphurein*
confused: *adiarthrôtos*
confute: *anairein*
conjoined: *sumphuês*
conjoined to/with, be: *sunaptein, suneinai*
connatural: *sumphutos*
consecutive: *ephexês*
consist in: *sunistanai*
consistent: *sumphônos*
consistent, be: *sumphônein*
conspicuously: *ekdêlos*
constitute: *sunistanai*
constrained: *biaios*
  by constraint: *bia, biâi*
construction: *oikodomêsis*
contact (n.): *haphê*
contact, be in: *plêsiazein*
contemplate: *theasthai, theôrein*
contend: *diateinesthai*
continuous: *sunekhês*
contract (v.): *sunairein*
contradictory (state): *antiphasis*
contrary: *enantios*
contrast (v.): *antidiairein*
contribute: *sunergein, suntelein*
contrive: *mêkhanasthai, paraskeuazein*
converse proposition (n.): *to antistrophon*
conversion: *epistrophê*
cool (v.): *psukhein*
cooperate: *sunergein*
co-present, be: *suneinai*
corporeal: *sômatikos*
corroborate: *bebaioun*
corruption: *phthora*
counter-argument: *antilogia*

counter-argument, make or give a: *antilegein*
counter-statement: *antirhêsis*
craft (n.): *tekhnê*
creation: *dêmiourgia*
creative cause: *paragousa aitia*
creator: *dêmiourgos*
crush: *piezein*

declaim: *rhapsôidein*
declare: *apophainein*
deliberate (v.): *bouleuein*
deliberation: *proairesis*
delimit: *diorizein*
demiurgic: *dêmiourgikos*
demonstrate: *apodeiknunai*
demonstration: *apodeixis*
denying: *apophasis*
depart: *ekbainein*
derive: *deiknunai, lambanein*
destruction: *phthora*
different: *diaphoros*
difficulty: *aporia*
difficulty, raise a: *aporein*
digestion: *pepsis*
dimension: *diastasis*
diminish: *meioun*
discourse (n.): *logos*
discover: *heuriskein*
discussion: *logos*
dismiss: *dialuein*
dispersed: *diespasmenos*
display (v.): *epideiknunai*
disposition: *diathesis, hexis*
dispositional: *kata tên hexin*
dissimilarity: *anomoiotês*
dissipate: *diaphorein*
dissipative: *skorpistikos*
distend: *diistasthai*
distinct: *diêirêmenos*
distinct, be: *diairesthai*
distinction: *diaphora*
distinctly: *diakekrimenôs*
distinguish: *dialambanein*
  be distinguished: *khôrizesthai*
divert: *apoklinein*
divide, make a division: *diairein, dialambanesthai, merizein*
divine (adj.): *theios*
division: *diairesis*
do: *poiein*
doctrine: *doxa*
doubt (v.): *diamphiballein*
down, downwards: *katô*
drag down: *kathelkein*
dragged, be: *helkesthai*
drink (n.): *poton*
dry (v.): *xêrainein*
dumb: *anoêtos*
duties: *(ta) kathêkonta*

eager to, be: *philoneikein*
earlier: *proteron*
earthy: *geêros*
element, elemental: *stoikheion*
end (n.): *telos, eskhatos*
endless: *ateleutetos*
enough, be: *arkein*
enquire: *zêtein*
enquiry: *zêtêsis*
ensouled: *empsukhos*
entire: *holos, pas*
environment: *to periekhon*
equal validity, with: *isotimôs*
equivalent, be: *isodunasthai*
establish: *kataskeuazein, sunistanai*
  also establish: *proskataskeuazein*
estrange: *allotrioun*
eternal: *aiônios*
everlasting: *aïdios*
everlastingness: *aïdiotês*
every: *pan*
evident: *dêlos, enargês, phaneros, prodêlos, prophanês*
  in an evident manner: *enargôs, phaneros*
evidently: *enargôs, dêlos, phaneros*
examination: *episkepsis*
exerting activity: *energêtikos*
exhalation: *anathumiasis*
exhale: *ekphusan*
exhaustion: *maransis*
exhibit: *paradeiknunai*
existence: *huparxis*
  existence, come into: *huphistasthai*
  existence to, give: *huphistanai*
  exists, what: *to on*
expand: *ekteinesthai*
expanded fashion, in an: *anêplômenôs*
explain: *apodidonai*
expose: *paragumnoun*
expulsive: *apokritikos*
extend: *ekteinein*
extension: *paratasis*
external: *exôthen*
extinction: *sbesis*

fail (v.): *epileipein*
fall (n.): *palaisma*
false: *pseudês*
fashion (v.): *dêmiourgein*

fashioning: *dêmiourgia*
fiction: *plasma*
fictional: *plasmatôdês*
field (n.): *pragmateia*
figure (n.): *skhêma*
fine, become: *leptunein*
finite, be: *perainesthai*
firm: *asphalês*
first: *proteros, prôtos, proton*
first place: *arkhê*
  in the first place: *tên arkhên*
fit (v.): *epharmottein*
follow: *akolouthein*
follow from: *sunagesthai*
food: *brosis, trophê*
force (n.): *bia*
foremost: *kuriôtatos*
form (n.): *eidos*
formation: *sustasis*
formulate: *eipein, poiein*
fulfilment: *apoplêrôsis*

general, in: *holôs, koinôs*
general manner, in a: *katholou*
generate: *gennan*
  be generated: *ginesthai*
generated: *genêtos*
generation: *genesis*
gifted: *euphuês*
go on to say: *epagein*
goal: *skopos*
god: *theos*
grasp (v.): *lambanein*
grow: *auxanesthai*
growth: *auxêsis*

habitus: *hexis*
happen: *sumbainein*
harm (v.): *lumainesthai*
heap up: *sôreuein*
heart: *kardia*
heat (v.): *thermainein*
heavens, the: *ouranos*
heavy: *barus*
here: *nun*
hinder: *empodizein, kôluein*
hindrance: *to kôluon*
  without hindrance: *anempodistôs*
hold (true): *ekhein*
honour (n.): *timê*
hypothetical syllogism, in a: *hupothetikôs*

identical: *hôsautôs*
ignorance: *agnoia*
ill, be: *kamnein*
imagination: *phantasia*
immobility: *akinêsia*
impart: *metadidonai*
imperishable: *aphthartos*
impermanent: *amenênos*
impose: *empoiein*
impossible: *adunatos*
improper sense, in an: *katakhrêstikôs*
impulse: *hormê*
in the same manner: *kata ta auta*
inanimate: *apsukhos*
incidental, incidentally, in an incidental manner: *kata sumbebêkos*
incompatible: *asunaktos*
incomplete: *atelês*
  less complete: *atelesteros*
incongruity: *allokotia*
increase (v.): *auxein*
increase (n.): *auxêsis*
increased: *auxêtos*
indicate: *dêloun, emphainein, endeiknunai, hupodeiknunai*
indisputable: *anamphisbêtos*
individual: *atomos, en merei*
inexhaustible: *adiexitêtos, anekleiptos*
infer: *metabainein, sunagein*
inference: *sunagôgê*
infinite: *apeiros*
infinity, at/to: *eis apeiron*
inopportune: *akairos*
inseparable: *akhôristos*
instill: *empoiein*
insult (v.): *paroinein*
intellect: *nous*
intellect, discursive: *dianoia*
intend: *boulesthai, epikheirein*
interpret: *exêgeisthai*
interpretation: *ekdokhê*
interpreter: *exêgêtês*
introduce: *epagein, paragein*
intuition: *epibolê*
invalid: *sathros*
investigate: *zêtein*
  that which is investigated: *to zêtoumenon*
investigation: *episkepsis, zêtêsis*
irrational: *alogos*
irregular: *plêmmelês*
irrelevantly: *para thuras*

join: *sunaptein*

know: *eidenai*
knowledge: *epistêmê*
  that which knows: *to epistêmon*

lacking nothing: *anelippês*
later: *husteron*
learn: *manthanein*
learning: *mathêma*
lengthen: *mêkunein*
lever (n.): *mokhlos*
licentiousness: *akolasia*
life, sign of: *to zôtikon*
light (adj.): *kouphos*
light body/constituent/entity/thing: *to kouphon*
lightness: *kouphotês*
limit (n.): *peras*
linked, be: *sunaptein*
little: *baios*
live (v.): *zên*
local: *topikos*
locomotion: *phora, topikê kinêsis, kata topon*
lower (adj.): *katô*

madness: *paranoia*
magnitude: *megethos*
maintain: *phulattein*
make a transition: *metabainein*
manifest (adj.): *ekphanês*
manifest (v.): *emphainein*
manifest, more: *prophanesteros*
manifestation: *ekphansis*
manifestly: *ekphanês*
manner: *tropos*
manuscript: *antigraphon*
matter, tied to: *hulikos*
  in a material way *hulikôs*
maturity, reach: *teleiousthai*
mean (v.): *sêmainein*
mean term: *mesotês*
mention (v.): *legein*
middle: *mesos*
mislead: *paralogizesthai*
mistaken, be: *planasthai*
mixed with, be: *sunkrinesthai*
moisten: *hugranein*
moment, at a certain: *pote*
motion: *kinêsis*
  capable of moving: *kinêtikos*
  perform the motion: *kineisthai, kinêsin*
motion, lack of: *akinêsia*
motive: *kinêtikos*
move (v.): *kinein, kineisthai, pheresthai*
  be moved: *kineisthai, pheresthai*
  can be moved: *to kinêtos*
  always moved: *aeikinêtos*
  moving: *kinêtikos*
move on: *elthein, metienai*
moved entity/part, that which is moved: *to kinoumenon, to kinêton, to kinêtikon*
movement: *kinêsis*
mover: *kinêtikon*

natural: *phusikos*
naturally: *phusikôs*
nature: *phusis*
  according to nature: *kata phusin*
  by/in nature *phusei*
  contrary to/against its nature: *para phusin*
  natural, naturally: *kata phusin*
  (be of/have a) nature: *pephukenai*
natural: *phusikos*
natural philosopher: *phusikos*
necessarily: *pantôs*
necessary: *anankê, anankaios*
  of necessity: *ex anankês*
note (v.): *ephistanai*
notice (v.): *ephistanein*
nourish: *trephein*
now: *nun*
numerically: *arithmos*

object (v.): *enistanai*
object to: *hupantan*
object of desire: *orektos*
objection: *enstasis*
obstacle: *empodion, to empodizon*
obstruction: *to huphistamenon*
obvious: *prokheiros, phaneros, prodêlos*
occur: *ginesthai, sumbainein*
offer (v.): *prospherein*
of the same kind: *homophulos*
oppose: *antidiatattein*
opposite: *antikeimenos*
opposite of, be: *antikeisthai*
order (v.): *suntattein*
order (n.): *taxis*
order, set in: *diakosmein*
outright: *autothen*
outside, from: *exôthen*
overcome: *katalambanein*
overturn: *anatrepein*
own (adj.): *oikeios*

paint: *diazôgraphein*
part (n.): *meros, morion*
partake: *metalambanein*
partake of: *metekhein*
particular: *merikos*
partless: *ameres* partlessness: *amerês*
pass over: *metakhorein, metienai, parêkein*
passage: *parodos*

passive: *pathetikos*
passive (component), the: *to paskhon*
  in a passive way: *pathêtikôs*
passively: *pathetikôs*
perfection: *teleiotês*
periodic return: *apokatastasis*
perish: *phtheiresthai*
perishable: *phthartos*
perishing: *phthora*
permutation: *exallagê*
permute: *exallatesthai*
perpetuity: *aidiotês*
per se: *kath' hauto*
pertain: *epiballein*
philosopher: *philosophos*
philosophy: *philosophia*
phrase: *lexis*
physical: *phusikos*
  in a more physical manner: *phusikôteron*
pile (n.): *surphetos*
place (n.): *topos*
  of place: *topikos*
plain, make: *apophainesthai*
posited, what has been: *to keimenon*
position (n.): *thesis*
possess: *ekhein*
possession: *hexis*
possible: *dunatos*
possible, be: *endekhesthai*
posterior: *to husteron*
potential, the: *to dunamei*
potentiality: *dunamis, to dunamei*
potentiality, in: *dunamei*
potentially: *dunamei*
power: *dunamis, dunasteia, exousia*
precede: *proêgeisthai*
preceding, immediately: *prosekhôs*
prefer: *prokrinein*
preponderant, be/become: *epikratein, pleonazein*
presence: *parousia*
present, be: *paradidonai, pareinai*
present, become: *enistanai*
present in, be: *huparkhein*
present itself: *anaphuesthai*
prevent: *kôluein*
previous: *proteros*
previously, *proteron, proteros*
pride oneself on: *philotimeisthai*
primary: *proêgeisthai*
principally: *proêgoumenôs*
principle: *arkhê*
prior: *proteros*
privation: *sterêsis*
probatory: *deiktikos*
problem: *skemma, problêma*
problematic: *aporos*
problematic, be: *aporeisthai*
proceed: *proienai*
produce (v.): *empoiein, poiein, paragein, proagein*
produced, be: *ginesthai*
producer: *to poioun*
productive: *oistikos, poiêtikos*
  that which produces action: *to poiêtikon*
profit (n.): *khrêmatismos*
progress (n.): *probainein*
progression: *proodos*
proof: *epikheirêma, pistis*
propel: *ôthein*
proper: *oikeios*
proper (characteristic): *idios*
property of: *to idion*
propose: *proballesthai*
proposed, be: *prokeisthai*
propound: *proballein*
prove: *deiknunai*
proximate: *prosekhês*
proximately: *prosekhôs*
put to the test: *basanizein*
puzzle (n.): *aporia*
puzzle, be/pose the: *aporein*
  present a puzzle: *aporian, parekhein*
  the puzzle is, *aporeitai*

quality: *to poion, poiotês*
  of such/this quality *poios*
quantity: *to poson*
  of this/such quantity: *posos*
question (n.): *zêtêsis*

raise: *ekhein, kinein*
raise puzzles: *aporein*
rank (n.): *taxis*
reading: *gegrammenon, graphê*
reality: *pragma*
reasonable: *eulogos*
  according to reasons: *kata logous*
  not reasonable: *alogos*
reasonable, it is; reasonably: *eikotôs*
receive: *dekhesthai, iskhein, lambanein*
  receive in addition: *proslambanein*
recommend: *keleuein*
reduce: *apagein*
reduction: *apagôgê*
refer to: *apoteinesthai, apoteinein*
referred back to, be: *anaphoran ekhein*
refutation: *elenkhos*
refute: *dialegein*
refute beforehand: *proanairein*

reject: *apodokimazein*
related to, be: *huparkhein*
relation: *skhesis*
relative: *pros ti*
reliable: *axiopistos*
remain: *menein*
remarkable way, in a: *thaumastôs*
remind: *hupomimnêskein*
remove (v.): *aphairein*
repeat oneself: *palillogein*
reply (v): *hupantan*
representation: *phantasia*
require: *axioun*
resolution: *dialusis*
resolve (v.): *luein, dialuein*
responsible: *aitios*
rest (n.): *êremia*
rest, be at: *êremein*
rest, cause: *êremizein*
rest, come to: *histanai*
rest, coming to: *stasis*
result (n.): *ergon*
retentive: *kathektikos*
return (v.): *epanastrephein*
return, periodic: *apokatastasis*
reveal: *ekphainein*
revolution, complete a: *apokathistasthai*
revolve: *perierkhesthai*
right, is: *kalôs*
rise (v.): *epanistanai*
rise to, give: *parekhein*
rise to the surface: *epipolazein*
run out: *epileipein*
rush (v.): *hormein*

sagacity: *ankhinoia*
sage (adj.): *spoudaios*
same: *homoios*
same kind, of the: *homophulos*
same manner, in the: *kata ta auta*
say: *eipein, legein, phanai, eresthai*
say, go on to: *epagein*
secrete: *enkrinein*
secretion: *enkrisis*
see: *horan*
seek: *zêtein*
seem: *dokein*
select (v.): *eklegein*
self-evidence: *enargeia*
self-moved: *autokinetos*
self-subsistent: *authupostatos*
sense (n.): *tropos*
sense (v.): *aisthanesthai*
sense perception: *aisthêsis*
sense perception, have: *aisthanesthai*
separate (v.): *diakrinein*
separate (adj.): *kekhôrismenos*
separated, be: *khôrizesthai*
separation: *diakrisis*
sequence (n.): *akolouthia*
set alongside: *paratithenai*
set forth: *ektithenai*
set in order: *diakosmein*
settle under: *huphistasthai*
shake (v.): *saleuein*
show (v.): *deiknunai*
show forth: *anaphainein*
similarly, (in a) similar way (or manner): *homoiôs*
  the case is similar: *homoiôs ekhein*
simple: *haplous*
simply: *haplôs*
size: *megethos*
sleep (n.): *hupnos*
sleep (v.): *katheudein*
solution: *lusis*
solve, give a solution: *luein*
sound (n.): *phthongos*
sound out: *perikrouein*
soul: *psukhê*
  without soul: *apsukhos*
speak: *eipein*
species: *eidos*
standing-still: *stasis*
star: *astêr*
stare (v.): *atenizein*
start (v.): *hormein*
state (v.): *legein*
state (n.): *sustasis*
step, as a next: *prosekhôs*
stone: *lithos*
stop (v.): *histanein, histasthai, pauein*
straight line: *eutheia*
straightforward: *antikrus*
straightforwardly: *antikrus*
strictly speaking, in the strict sense: *kuriôs*
strike: *krouein*
string: *khordê*
stupid: *anoêtos*
submerged, be: *buthizesthai*
submit: *hupomenein*
subsist together: *sunupostasis*
subsistence: *hupostasis*
substance: *ousia*
substantial: *ousiôdês, kata genesin*
substantial, less: *kouphos*
subsume: *hupotattein*
succeed: *diadekhesthai*
sufficient, be: *arkein*
sufficient: *autarkês*

suggest: *hupomimnêskein*
suitability: *epitêdeiotês*
summit: *akron*
superiority: *huperokhê*
supernatural: *huperphuês*
support (v.): *anekhein*
syllogism: *sullogismos*

take into account: *prologizesthai*
take into consideration: *theasthai*
take place: *ginesthai*
take together: *sunairein*
talk (v.): *legein*
temperament: *krasis*
tend to: *pephukenai*
term (n.): *horos, onoma*
  use the term: *onomazein*
test, put to the: *basanizein*
theory: *theôrêma*
think: *dokein, oiesthai, noein*
thought, discursive: *dianoia*
thought: *noêma*
through the mediation of: *mesos*
throw in: *paremballein*
time: *khronos*
tool: *organon*
topics, present: *ta prokeimena*
touch (v.): *ephaptesthai, haptesthai*
transformation: *metabolê, metastasis*
transition: *metabasis*
transition, make a: *metabainein*
transitionally: *metabatikôs*
transported, what can be: *phorêton*
travel: *hodeuein*
treatise: *pragmateia*
trip up: *sphallein*
true *alêthês*
truth: *alêtheia*
twofold: *ditton*

up, upper, upwards: *anô*
unceasing: *adialeiptos*
unclear: *adêlos*
unconvincing: *adokimos*
undemonstrated: *anapodeiktos*
understand: *akouein, apodekhesthai, ekdekhesthai, parakolouthein*
understanding: *ekdokhê*
undoubted: *anamphilektos*
unengendered: *agenêtos*
unextended: *adiastatos*
unfailing: *anekleiptos*
ungenerated: *agenêtos*
unhindered: *anenokhlêtos*
uniform stuffs: *homoiomereiai*
uninterrupted: *anekleiptos, adiakopos*
universally: *kathalou*
universe, the: *to pan*
unmoved: *akinêtos*
unsound: *sathros*
unwind: *ekmêruein*

vain, in: *matên*
verification: *epikrisis*

walk (v.): *badizein*
warm (adj.): *thermos*
warm (v.): *thermainein*
  entity which is warmed: *to thermainomenon*
warmth: *thermotês*
waste away: *aporrhein*
way: *tropos*
weight: *baros*
what follows: *to hepomenon, ta hexês*
whole: *holos*
wish (v.): *boulesthai*
with equal validity: *isotimôs*
withdraw: *aphistasthai*
withdrawal: *anakhôrêsis*
within, from: *endothen*
without: *khôris*
without, from: *exôthen*
without further qualification: *haplôs*
witness, bear: *marturein*
world: *kosmos*
worth *axios*
worthy: *axios*
write: *graphein*

Greek-English Index

'Other tr[anslation]' or 'paraphrase used' means there is no easy one-to-one correspondence between the Greek and the English.

Subject Index

Roman numerals and one or two digit Arabic numerals (e.g. viii or 13) refer to the page numbers of the introductory material, numbers of the form 1232,24 refer to page and line numbers of the Greek text translated here, which are printed in the margins of the translation. Numbers prefixed n. or nn. refer to notes, those preceded by a page number from the introductory material (e.g. ix n.2 or 14 n.19) referring to the notes to that material, the rest to the notes to the translation.

www.ingramcontent.com/pod-product-compliance
Ingram Content Group UK Ltd.
Pitfield, Milton Keynes, MK11 3LW, UK
UKHW020138290926
15471UKWH00003B/562

9 781472 539175